CONTENTS

PREFACE

		Page
I	Overview	*1*
II	The Underlying Philosophy	*2*
III	The Implementing Structure	*2*
IV	Notes to the Instructor	*3*
V	Notes to the Student	*6*
VI	Acknowledgements	*10*

EXPERIMENTS*

0	Getting Started – Instruments and Measurment	1
1	Operational-Amplifiers Basics and Beyond	13
2	Operational Amplifiers Imperfections and Applications	23
3	Junction-Diode Basics	29
4	Bipolar-Transistor Basics	35
5	MOSFET Measurement and Applications	43
6	The BJT Differential Pair and Applications	51
7	Single-BJT Amplifiers at Low and High Frequencies	61
8	Feedback Principles Using an Op-Amp Building Block	71
9	Basic Output-Stage Topologies	83
10	CMOS Op Amps	93
11	Op-Amp-RC Filter Topologies	101
12	Waveform Generators	115
13	CMOS Logic Characterization	125
14	TTL Logic Characterization	137

APPENDIXES

Contents Overview

A	Experimentation	149

The Role of Laboratory Experimentation; Laboratory Insights; Experiment Layout; Testing; Troubleshooting; Safety.

---

*Experiments, except for #0 and #1, cover the same material as the correspondingly numbered Chapter in the Text, "Microelctronics Circuits", 4/e, Oxford University Press, by Sedra and Smith.

B	**Instrumentation**	**163**

Power Supplies and Current Limiting; Oscilloscope Calibration; Oscilloscope Measurement; Differential Measurement; AC Measurement; Timing Measurement; Linearity and Distortion; Temperature Testing; Roles for Potentiometers; Roles for Resistors; Control of Parasitic Oscillation; Standard Component Values.

C	**Reporting**	**181**

The Role of Engineering Reports; Report Design as Preparation; Engineering Record Keeping; Report Formats; Standard Forms and Graphs.

Laboratory Explorations

for

Microelectronic Circuits FOURTH EDITION
Sedra / Smith

Kenneth C. Smith
University of Toronto
Hong Kong University of Science

New York   Oxford
OXFORD UNIVERSITY PRESS
1998

Oxford University Press

Oxford   New York
Athens   Auckland   Bangkok   Bogota   Bombay   Buenos Aires
Calcutta   Cape Town   Dar es Salaam   Delhi   Florence   Hong Kong
Istanbul   Karachi   Kuala Lumpur   Madras   Madrid   Melbourne
Mexico City   Nairobi   Paris   Singapore   Taipei   Tokyo   Toronto   Warsaw

and associated companies in
Berlin   Ibadan

Copyright © 1998, 1991 by Oxford University Press, Inc.

Published by Oxford University Press, Inc.,
198 Madison Avenue, New York, New York, 10016
http://www.oup-usa.org
1-800-334-4249

Oxford is a registered trademark of Oxford University Press

All rights reserved. No part of this publication may be reproduced,
stored in a retrieval system, or transmitted, in any form or by any means,
electronic, mechanical, photocopying, recording, or otherwise,
without the prior permission of Oxford University Press.

ISBN   0-19-511772-7

9 8 7 6 5 4 3 2 1

Printed in the United States of America
on acid-free paper

Cover Illustration: The chip shown is the ADXL-50 surface-micromachined accelerometer. For the first time, sensor and signal conditioning are combined on a single monolithic chip. In its earliest application, it was a key factor in the improved reliability and reduced cost of modern automotive airbag systems. Photo reprinted with permission of Analog Devices, Inc.

PREFACE

I OVERVIEW

- **The Manual Format**

    Laboratory Explorations is a **Manual** of *Laboratory Instructions and Aids* written in close support of the **Text** *Microelectronic Circuits, Fourth Edition*, by Sedra and Smith. As such, it relies heavily on that **Text** for context, background and theoretical support. Thus, here in the "**Manual**", the emphasis is more, so to speak, on archeology than on the opening of new lands, much more on the excitement of what is, than on the mystery of what might be. But, of course, just as history can illuminate the future, The Manual intends to prepare the student with a set of universal tools for the enlightening process of *learning by doing*!

    A total of 15 **Experiment**s are provided, about one per *Chapter* of the **Text**, there being none for *Chapter 1*. In general, each **Experiment** is a relatively basic one intended to support the most fundamental part of its associated **Chapter**. Accordingly, it normally deals most with the Chapter's the early part, although, occasionally, later material is also exposed. Experiments #2 through #14 cover corresponding Chapters #2 through #14. Experiments #1 and #2, both on op amps, support Chapter 2 of the Text. Experiment #0 stands alone, a basic one which exposes the student to the basic instrumentation which supports the remaining Experiments. In general, there is no explicit dependence of any **Experiment** on any earlier one, beyond an expectation of increasing maturity, awareness, and facility with measurement as the student progresses; As well, there is a selective degree of review, through the process of performing a relatively basic task initially, one which may have appeared before, as the preamble to a more profound experience.

- **The Experiment Format**

    Within each **Experiment**, there is a repeated high-level structure embodied in the Section labels: *Objectives, Components and Instrumentation, Reading, Preparation and Explorations*. For the choice of these words, many of the motives should be obvious: **Objectives** attempts to motivate, intrigue, and foreshadow; **Components and Instrumentation** provides a context, with reference to supporting material in the **Appendix**es, of which more is said, later; **Reading** is a simple statement of the relevant parts of the **Text** *Microelectronic Circuits, Fourth Edition*, by Sedra and Smith, hereinafter, usually called the **Text**; **Preparation** is a collection of tasks and tests very closely coupled to the "**Explorations**", involving verification of a design, alternative designs, creation of hypotheses and expectations to be verified by experiment, and so on; **Explorations** is the heart of the process to which the rest aims, and of which much more will be said shortly;

    As noted, **Explorations** is the heart of each **Experiment**, and is presented as a multilevel hierarchy: In general, there are *parts* having a general theme, *divisions* emphasizing aspects of that theme and to which the collective term Exploration is often applied, and *tasks* intended to motivate investigations of that theme. While most of the tasks simply direct some measurement or observation, some request minor calculations to be done on the spot, whose outcome is expected to motivate, and possibly reorient, the exploration process itself. Periodically, occasionally at the end of a part, but primarily at the end of a division, yet often interspersed within a division, another level of reflection is motivated by the heading, **Analysis**, commencing with a preamble, **Consider**. The latter typically presents a global conjecture concerning the nature of the results, intended to motivate reflection on *what really happened*. Often explicit computations and comparisons are suggested as well. A usual preface to **Analysis** is an item labelled **Tabulation** whose role it is to attempt to capture what has been measured in the preceeding part, and to do so in a format which encourages the use of a **Table** (see Appendix C).

    Tabulation is intended to provide a somewhat structured sketch of items around which to organize a **Report**. This is provided in the context of two kinds of material presented in the relevant **Appendix**es (A, C), which include a generalized structure of a formal **report**, as well as a set of tables and graphs (available to be photocopied) as prototypes of information-compressing presentation formats. These particular **Appendix**es exemplify the spirit in which a relative-large amount of additional material is supplied to the student: one which emphasizes the opportunity to expand, rather than limit, horizons.

In fact, the **Appendix**es, as a whole, are quite broad-ranging as the table of **Contents** attests. In general, they provide motivation, background and advice, as well as information on items for which there is a recurring need. Throughout the body of a typical **Experiment**, there are continuing references to specific **Appendix**es and to Experiment #0 as memory joggers to the possibly unaware. The format is thereby intended to be thorough and complete, but **not** repetitious, space-wasteful, or boring.

II THE UNDERLYING PHILOSOPHY

The question "*What is the ideal format for the presentation of laboratory instructions?*", while easily asked, typically provides answers not easily used. The difficulty seems to lie, literally, in personal and private aspects of the situation: While almost no one likes to be told what to do, many *need* such direction, whether or not they realize or acknowledge it! While most respondents want the instructions to be clear, many bridle at them being overly detailed! While most want them complete, few want them tedious! And yet the meaning of all of these measures of quality is coloured unconsciously by the background and knowledge of the respondent! And *that* is a major problem for us all!

All of this is why most Instructors would prefer to prepare their *own* Instruction Sheets for their *own* laboratories, if time would only permit! For they know themselves to be in the best position to know what the student needs, and what the student has the background to handle. But what does the student think? We know what she says; but what does he mean?

It was the search for a way around this morass, which led to the basic form of the first edition of the **Manual** and its underlying philosophy of **Explorations**. The general idea embodied in the latter word is one of travel through unfamiliar territory, providing an opportunity to poke and probe and to discover, for oneself, the hidden "treasure". Yet to arrive expeditiously at the territory of greatest interest, every explorer needs a map. No matter how familiar she/he is with the general territory, the map can make the trip to the edge of the less-known land, a more efficient one.

Thus, in the first edition of this **Manual**, a role was seen for the existence (but not necessarily the enforced use) of detailed and *procedural instruction* as a *key element* in making progress more efficient and interesting. Not just "go downtown and see the sights" but, rather "take the #9 bus south on Main St. and notice the new structure at Cross St., the activity at the square, etc, etc". Now we all know that you may hate the bus, dislike modern architecture, and abhor crowds, in which case, you avoid the hassle, and use your car! And so it is with the process of exploration in these **Experiments**. You *need not follow* the detailed instruction, *but if you want to, or if you need to*, it is there!

The present Manual, the second edition, of *Laboratory Explorations* maintains much of this general philosophy, but with several levels of simplification in detail. Besides being smaller, its average **Experiment** is more basic, closer to the level of necessity than many in the first edition. Moreover, individual **Experiments** are reduced in complexity and shortened. Relatively few optional tasks are discussed explicitly. However, some **Experiments**, particularly those for more advanced chapters remain long, simply because there are lots of possible interesting things to do, and no obvious way to pick a "best" subset in view of the diverse possible curricula these Chapters of the **Text** support.

III THE IMPLEMENTING STRUCTURE

- **The Procedural Hierarchy**

Each **Experiment** in this **Manual** is presented in a relatively consistent pattern, a pattern which is intended to make the selection of your **Experiment** itinerary easier, and possibly more dynamic:

Specifically, in each **Experiment**, there are five **Sections** labelled: I **Objectives**; II **Components and Instrumentation**; III **Reading**; IV **Preparation**; and V **Explorations**. Section V, **Explorations**, is the primary one to/from which all others are directed. Consequently, most **Figures** are embodied within **Explorations**, being placed in close proximity to the instructions which concern them. The exception is for **Figures** which provide component-pin connections, that are located earlier, in Section II. More importantly, **Preparation** follows the same overall gross-division pattern originating in **Explorations**, using distinguishing prefixes P and E respectively.

Signposts indicating the general direction of the **Experiment** are provided at four levels: At the *first* level, overall direction is presented in Section I **Objectives**, which contains a very broad view of the journey, plus some comment on its general context, both within this **Manual** and in the World at Large. At the *second* level, more explicit, yet broad, direction is provided within Section V **Explorations**, in the large bold-face title and their preamble. At the *third* level, entries in each **Exploration**, labelled **Goal** serve in more detail to project the direction. Finally, to round off the **Exploration**, summarize one its major parts, is the entry: **Analysis**, where items to **Consider**, present often in explicit detail, an encapsulation of the main points of interest, while hinting at interesting new vistas, and (ideally) motivating new exploratory excursions. Overall, these signposts are intended to help in several ways, some obvious, and some less so. Thus, for example, if you wish to get an idea of the journey ahead, and where you are going in general, it is a good idea to read these signs early: In practice, an initial preview of the whole **Experiment** using them is highly recommended.

- **Reporting**

Reporting, as the one immediately tangible result of a laboratory experience, is obviously an important issue, but another one on which differences of opinion can exist, as noted later in this **Preface**, Part IV *Notes to the Instructor*. While, as indicated there, the reason for reports may be a topic of discussion and concern, there is little or no question about the underlying nature of the **Report,** whether formal, or otherwise. Without doubt, independent of its degree of formalism, a laboratory report must be clear, concise, correct, and as complete as the time budget allows. It is to these ends that parts of each **Exploration**, called **Tabulation**, and **Analysis**, as well as **Appendix C** are directed. The intents are several: *First*, **Tabulation** provides a concise summary of virtually all that has been requested as raw measurement data in the **Measurement** steps of the **Exploration**. *Second*, **Analysis** specifies expected derived results. *Third*, by the use of the very word **Tabulation** attempts to reinforce the idea that preplanning of the final presentation format can make the actual measurement process much more efficient, following the old adage, "A place for everything, and everything in its place". (There is much said in general about reports and reporting, and more specifically about Tables, in Appendix C.)

However, by design, there is no explicit instruction for the final reporting format, whether formal of not. That is left up to the Instructor, with guidance and encouragement provided for the use of the student, as needed, in Appendix C.

IV NOTES TO THE INSTRUCTOR

- **Choosing the Experiment**

The material in this **Manual** is appropriate for a wide variety of instructional styles, in a sequence of two or three single-semester courses in electronics but with some focus on the introductory course. While the overall emphasis is on exploration and adventure, the material is intentionally structured, through its embedment in a procedural shell, intended to allow for a spectrum of student skills, possibly with a minimum of immediate instructional assistance, such as in an open-laboratory setting.

Relative to the three-part structure of the **Text** "*Microelectronic Circuits*", *Fourth Edition*, by Sedra and Smith, the organization is as follows: There are 6 **Experiments** (#0 and #1 through #5) explicitly for Part I, on Devices and Basic Circuits; 7 (#6 through #12) explicitly for Part II, on Analog Circuits; and 2 (#13 through #14) for Part III, on Digital Circuits. Experiment #0 is intended to provide a general introduction to laboratory instruments and appropriate measurement technqiue.

Relative to the course structure presented on pages *x* and *xi* of the Text, the potential choice is as follows: In the first course, for *Possibility* (a), there are 5 or 6 relevant **Experiments** (#0 and #1 through #5, perhaps excluding #2); for *Possibility* (b), there are 6 relevant **Experiments** (#0 and #3 through #7), for

Possibility (c), there are 8 relevant **Experiments** (#0 and #3 through #7, selected parts of #13 and #14). In the second course, for *Possibility* (a), there are 7 relevant **Experiments** (#6 through #12); and for *Possibility* (b), there are 5 relevant **Experiments** (#6, #7, #8, #13 and #14).

- ### Assessing the Degree of Difficulty

In general, the length and difficulty of each Experiments has been adjusted according to the expected degree of maturity of the learning student and the degree to which the material is a variable element in curriculum and course design. Thus, **Experimen** #0 on basic laboratory practice, is a special case: While long, it is not particularly difficult, and certainly optional. **Experiment** #1 through #7 and #9 are intentionally reduced to emphasise relatively basic material. In general, there is an increase in length and possible need for selection of areas of concentration in Experiments #11 through #14. Experiment #10 on CMOS Op Amps is quite difficult and long. Experiment #11 on filters is long and multifacetted, as in #12, but to a lesser extent. Finally #13 and #14, on MOS and BJT digital topics respectively are quite long, with parts ranging from the simple to the complex, in an attempt to meet various needs.

Of course, besides the possibility of simply doing everything in turn, there is the obvious opportunity for selecting only a few of the **Explorations** in an **Experiment**. The former linear process is appropriate only for a course of study in which depth of understanding is emphasized, while nominal completion in a fixed time, is not. It is suited primarily for open laboratories, and can be appropriate as well, for situations in which formal reporting is de-emphasized[1] in favour of real-time understanding.

One final comment, related to the apparent *size* and *complexity* of an **Experiment** is in order. Simply put, it is apparent that these measures are highly subjective, depending obviously on many things: on the background of the student, on his degree of familiarity with the topic, on her familiarity with experimentation in general, on her/his manual skills, in short, on the relevant "experience" the student brings to bear. Thus, *any* experiment for *any* individual can be more or less difficult, or more or less trivial, more or less boring, etc. But bear in mind that these problems imply opportunities as well: For, occasionally, to try to do something that is beyond one's immediate experience is mind-expanding, and to redo thoroughly and intensely what one already "knows" can be knowledge consolidating, all, of course, if and only if the stage is set appropriately by you, the Instructor.

- ### Choosing Other Options

Beyond the straightforward use of the material presented through explicit selection of divisions and tasks, other possibilities exist:

- **Informal Options**: Consider structuring related and/or, alternative experimental paths, based simply on a combination of the **Explorations**, *Title*, *Goal* associated *Figures*, and *Consider* entries. In this approach, the **Considers**, in conjunction with the **Figures**, would be viewed as setting the stage for a more spontaneous, barely-structured, experimental experience, for which the main body of the **Explorations** would be resource material, to be used for help or inspiration, as needed.
- **Simulation Options**: There are many possibilities of introducing simulation into these **Experiments**. One could consider PSpice for specific circuit simulation, or Electronics Workbench[2] for a somewhat complete experience, as either a supplement or alternative to physical experimentation. These include:
    - Simulation as an adjunct to the suggested **Preparation**.

---

[1] See *Reduced-Report Options* and, subsequently, *Choosing the Reporting and Experimentation Style* to follow.

[2] "Electronics Workbench", Interactive Image Technologies, Ltd. Toronto, Ontario, Canada is an interactive laboratory simulation tool, running under Windows 95. A CD-ROM sampler including "Workbench" is provided with "Microelectronic Circuits", 4/e, Oxford University Press, by Sedra and Smith. A full set of the circuits in this **Manual** are being prepared for "Workbench" simulation, will be made available at a later date.

- Simulation as an alternative to the suggested **Preparation**.
- "Electronics Workbench" as an alternative (or adjunct) to hands-on **Experimentation**.
- A laboratory-session pattern in which alternating laboratory sessions are devoted to preparatory simulation and physical experimentation.

- **Reduced-Report Options**: It is worthwhile to consider the degree and method of reporting as a control variable for the regulation of student pace. Many variants are available:
  - With two-person parties, both members pre-prepare tables, then one measures and dictates, while the other records.[3]
  - Select parts to be done, but for which *little* or **no** *reporting* is required (perhaps, beyond a simple checkoff).
  - Prescribe a bare report with raw data *only* (with no commentary), as the *immediate* product of the Laboratory experience, with a *later* requirement for a formal report, possibly assigned, or of the student's choice, or chosen by lot.
  - Prescribe a summary report in cursive style motivated only by the **Consider**s (possibly, as an option, to be to done within the Laboratory period).

Reflect on the likelihood that to "walk in the woods and smell the flowers" is to enjoy, to appreciate and to establish lifelong memories, while to traverse the woods in search of every flower and to document its exact location is to spoil both the experience and the landscape, and all of that for information nearly trifling and certainly ephemeral! Is our goal to educate or to train? Clearly it is both, but the which, the where, the when, and the wherein must be considered carefully!

• Choosing the Equipment

With respect to the equipment employed in these **Experiment**s, the underlying premise has been that of a relatively cash-starved undergraduate teaching laboratory. Accordingly, the **Experiment**s are based on a modest set of essential equipment, namely two power supplies, a DVM a two-channel oscilloscope, and a waveform generator. In this sense, the attempt is to demonstrate to the student that cunning and ingenuity are valid substitutes for wealth, however unglamorous that substitution may be.

Of course, more equipment would be more convenient. A second DVM, a pulse generator, then a third supply would have highest priority. As well, a capacitance meter would be quite handy and informative. Otherwise a larger shareable, more precise RLC bridge would suffice. Clearly, a shareable component-curve tracer would be handy, and a spectrum distortion/waveform analyzer would be very nice! And so on ⋯.

Note, that the nature of the equipment available can evidently affect the possible complexity and sophistication of the individual **Experiment**. Thus, for example, while such things are only rarely mentioned in the **Experiment**s, the availability of a distortion analyzer would clearly enrich *all* of the amplifier experimental investigations; likewise, access to a curve tracer, even broadly-shared, would both extend the scope of comparative device measurements, as well as reduce the tedium and timescale of some of the procedures. It is apparent that you, the Instructor, alone can factor these issues into the material which constitutes each **Experiment**. Generally speaking, since the bulk of the instruction provided is on the setup of the circuit itself, it would seem that augmentation with more sophisticated measuring equipment would constitute a relatively small change, yet possibly one, if implemented, could reduce the time, and/or reorient the direction of the entire **Exploration**, quite dramatically.

• Choosing the Reporting and Experimentation Styles

As noted many times already, the amount of material in some **Experiment**s can appear to be quite large and potentially overwhelming. Yet this view itself is highly context-laden, dependent, for example, on one's assumption of reporting style. Without question, to maintain a well-documented trail of results for even some

---

[3] This, of course, is a mode often adopted by students themselves, but it could usefully be formalized.

parts of some **Explorations** can be a very large and time-consuming task. Yet such is necessary to prepare a thorough formal report.

But, why, in fact, do we traditionally ask for thorough reports? One immediate answer, of course, is simply to state that to be an engineer is to be thorough; it is simply our responsibility! Yet we as educators have other immediate goals – to encourage understanding of the nature of the world, for example. It is toward such greater goals that traditional formal reporting of work done in a teaching laboratory can be counter-productive, if formal reporting is *always* emphasized to the exclusion of other approaches. For often what we require from the student in response to his Laboratory experience is *only* a formal presentation, and we receive such documents, and we grade such documents, but what do they represent? Where do they originate? The fact is that the formal laboratory report, conceived as a means to an end, has become the end itself! Usually we assume that a good report characterizes the state of mind of an effective student. Yet we know there are other possibilities ···!

Let us face the fact that formal reports, though important in context, are *not all* that Laboratories are about. Rather, learning and understanding are the *real* issues. Reports are merely vehicles for a somewhat questionable evaluation process!

Whether you agree or not with all of this, one fact remains: It is simply that formal reports, well and thoroughly done, take a lot of time, time which could be spent, in part, in other ways. Are there ways which may more quickly lead to true understanding? Probably! One suggestion of relevance here, with the tendency toward a surplus of material this **Manual** presents in some of its structures, is to try to encourage less-formal, but more rapid, investigations, akin to the process of exploration without map-making, to which I alluded earlier.

Following the intent of this suggestion, a student could conceivably learn more to observe, to rapidly sense the direction in which the **Experiment** leads, while aware of, but not trapped in, its twists and turns. Metaphorically speaking, the quasi-static process of point-by-point plotting could be replaced by the dynamic one of curve tracing. As viewed in this "sweeping" way, many of the lengthy **Experiments** may become considerably more manageable.

But how, then, does the student know and recall what she/he has done? How do **we** know, as well? How do we evaluate the acquired state of mind? While alternative suggestions abound, this is not their place. Suffice it to say: • do not abandon formal reports entirely; • consider sampled reports of whole **Experiments** or subsections; • consider written tests in which informal laboratory notebooks are permitted; • consider actual laboratory timed-task tests; • consider two-phase (two-period) laboratory experiences – the first fast one to survey the field, the second painstaking one to examine its nooks and crannies. The latter is one possible way to look at an alternating pattern of simulated and physical experimentation sessions.

V NOTES TO THE STUDENT

• Coping

There is a very real possibility that the presentation format and the amount of material in some of the **Experiments**, particularly the later ones, will seem overwhelming to you: Certainly, *there is often a lot of material*! Certainly, the number of incremental steps presented to you is often large; But, there are also a lot of mechanisms in place to assist you. Your challenge is to identify them early, and to use them. Bear in mind that the presentation is for a very wide range of student needs and backgrounds. Ideally, your Instructor has set the stage for *you*, both in preparing *you* and selecting an appropriate program of experimental work for *you*. But we all know, you, your instructor, and I, that the match will never be perfect – something or another will be missed, by you or by us, either in the past, or in the present, something will be unclear and confusing. It is to help in this situation that these paragraphs are written especially for *you*.

The challenge ahead is very real, both specifically and in general. Laboratories are about reality, and the reality is that "*Life itself is a Laboratory*", a place to experience the old and the new, to poke and probe, to discover, to learn, and to *enjoy*! The bottom line is that in each, both Life and Laboratory, you must learn to help yourself *to be selective, to make decisions, to take action*. This is really what an Electronics Laboratory is ultimately all about. Though I hesitate, in this public context, to state it, Electronics, a topic which I dearly love, is nevertheless *only a vehicle* for carrying this message. Yes, a very *important* message, but also, of course, a very, very exciting vehicle!

• Your Support – This Manual

Try to be aware of, and keep in mind, the many elements in this **Manual** that are here to help, to help you, the Student of Electronics. First and foremost, strangely enough, is the **Appendix**. It is strange in the sense that students often see the **Appendix** for what it often is – a collection of random items for which there is no other place – a mere appendage. Yet, here, the **Appendix** can actually be viewed as the core around which the collected **Experiments** on diverse topics attain their unity. But where else can one put such material to be found easily and accessed? Clearly, either here in the **Preface**, where you now read, or *at the end*, in the **Appendices**.

I strongly recommend that you browse through the **Appendices**[4] soon, and read some of its parts, particularly the early ones, as soon as you have some spare time.

Lest you forget what you find there, in an **Appendix**, you will note, in the main body of the **Manual**, within each **Experiment**, on-going references to specific **Appendix**es, serving to remind you of this supporting material, while maintaining the theme flow.

But there are other mechanisms in each **Experiment** to help as well. Of these, the overall structure in **Sections** I through V, is perhaps the most in evidence: Section I **Objectives**, and Section III **Reading** are obviously important in establishing the general direction and context of the **Experiment** in your mind. Section II **Components and Instrumentation** describes attributes of the physical context in which you will work; Section IV **Preparation** has the role of directly focusing your attention on the work ahead, with forward reference to the **Figure**s and associated instructions. Section V **Explorations** is really the main body of each **Experiment**, around which the rest hovers in support. Generally speaking, a relatively large number of **Figure**s is intended to make the instructions more concrete and effective for you. To assist you, the **Figure**s are generally placed quite near their first point of reference, in order to minimize page turning as you proceed in the heat of your battle with the electrons.

Four other mechanisms exist as well in the body of V **Explorations**, for your use and assistance: Most obviously, there is often a tone-setting preamble associated with each major *Division* (the component with the BIG title). In the same vein, there is often a comment at the front end of each *Subdivision* (labelled **En.m**), and always a **Goal** whose attempt is to prepare for you a view of the horizon to which you are aimed. Then, using a mangled metaphor from your course in structured software, for each "Begin", there is an "End". This is in the form of the **Analysis** part with its **Consider** items, whose role it is to crystallize in your mind what the intervening steps were all about. As well, along the way, there are **Notes**, often about issues of safety, subtlety, or relationship, which, though very local in appearance, convey a sense of the nature of the skirmish underway.

• A Structure for Laboratory Learning

While there are a great many ways to proceed with any of the **Experiment**s presented here, some are much more effective than others. In particular, perhaps surprisingly, the straightforward approach of starting at the beginning and working to the end, is *a bad idea* for most students, under the circumstances most students face. This property is not unique to the particular **Experiment** format here; it is quite universally true of any structured document whose goal is to present a complete picture in an orderly fashion. If the reader has lots of time, the straightforward approach works quite well. Thus one normally reads a novel from beginning to end, primarily because one's reading of it has no enforced timescale: One simply wouldn't be reading it, if hard-pressed for time! Moreover, the novel format is essentially linear in its presentation. It is only the successful novelist who is able to use a book's enforced one-dimensionality to create effective images, both

---

[4] To help, the Table of Contents outlines the entire Appendix content. As well, each of the major Appendix divisions (A through C), includes a preview of its message at the top of its first page.

multidimensional in space and time. In this task, he employs an intricate juxtaposition of facts, statements and ideas, written serially, to be read serially, and virtually impossible to randomly access. Now to return to the issue at hand, this Manual, like any effective text, such as *Microelectronics Circuits*, by Sedra and Smith, for example, allows for, and even encourages, random access. For this is one of the techniques which makes such a book popular and leads to its acceptance by a broad audience. Thus, to repeat, but also to generalize, *do not proceed in any task of technical reading from beginning to end*, **unless you have a lot of time**. If time-constrained *in any way*, please employ a more selective and cyclic process. Don't just slog through the swamp in hope of reaching the end, climb a tree and discover where to go! Such a process is now described for your task at hand:

Preparing for a Laboratory Exploration:

A. Be aware of your Instructor's special directions;

B. Mark assigned parts in *your* **Manual** in **Preparation** and **Exploration**;

C. For the *assigned parts*:
- Read, carefully, the **Objectives**;
- Read, quickly, the **Titles** of the **Explorations** items;
- Read quickly, the **Goals**;
- Note, quickly, the associated **Figure**(s);
- Read, very quickly, the associated **Analysis** and **Consider** items;
- Read, very, very quickly, the **Preparation** items.

D. For each assigned **Exploration** (not *all*, but *each*, in turn):
- Read, carefully, the **Exploration** with **Goals** and instructions, examining the associated **Figure**(s), while attempting to get a mental picture of what is asked, noting requested items in the margin of the **Manual**, or on the **Figure**(s), by highlighting, etc;
- Read, carefully, the associated **Tabulation** items, checking against the **Exploration** instruction and your own earlier conjectures, and also organizing and preparing appropriate tabular forms for your Laboratory Notebook; [See **Appendix C**];
- Prepare, carefully, in your notebooks, additional circuit schematics, additional labelling of existing **Figure**(s) (of pins, etc), and possibly a sketch of the prototyping-board layout; [See **Appendix A**];
- Read, very carefully, the associated **Preparation**, sketching an answer as you go;
- Possibly, complete the associated **Preparation**, *or* proceed to repeat all of this (in the list labelled **D**) for the next assigned **Exploration**.

E. Finish all **Preparation**s not already complete;

F. Ensure that all labelled tables, diagrams, graphs, etc, are organized in an orderly way in your notebook, for all assigned **Explorations**;

G. Possibly, prepare component parts and wiring layouts for your prototyping board, if not already done;

H. Possibly, wire your circuits on your prototyping board prior to going to the Laboratory itself!.

After reading **all** of the above, your immediate thought may be "Is he nuts?!". Well, frankly, the answer is unclear! But what **is** obvious, is that what has been stated **is what should be done** in an ideal world. In fact, it is the metaphorical equivalent of **what must be done** by an effective engineer employed successfully in research, in development, in business, or in any other venture into the unknown!

But, obviously, reality must prevail, "There is so much to do, and so little time". So do what you can; ignore what you must! Bear in mind that work in a Laboratory, probably far more than any other activity short

But, obviously, reality must prevail, "There is so much to do, and so little time". So do what you can; ignore what you must! Bear in mind that work in a Laboratory, probably far more than any other activity short of going to the moon, benefits greatly from **Preparation**. For it has been said, wisely, that "Luck is the Encounter of Preparation with Opportunity".

• Overall Survival Tips for Laboratory Students

The Environment for Exploration

Laboratory Exploration should be an efficient process! In it, your actions should be "natural", and spontaneous. For this purpose, it helps to establish some "norms", ways of doing things which are consistent with effective engineering practice. Here are some **tips**.

1) Become very familiar with the instruments available to you. Know what each does in considerable detail. Normally, you will use a dual-channel oscilloscope with probes, a DMM, two power supplies, and a signal generator. You want to become very efficient at measurement. The best way is to practice a lot with a limited but effective tool set.[5]

2) Use a flexible "prototyping board" (the board, or the PB) with lots of connecting wires of appropriate lengths and different colours. You want connections to be quick, easy, reliable and traceable.

3) Use somewhat formal wires and cables to connect the power supplies and signal generator to your board. You do not want these basic connections to be unreliable or intermittent.

4) Use "unit values" of resistors and capacitors (such as 1 kΩ, 10 kΩ, 10 μF, 0.01 μF, etc.) as much as possible. You want component selection to be quick and easy, and "order-of-magnitude" (mental) calculation to be very convenient and effective.

5) It is best and most efficient for you to have a generous supply of components and wires, organized by size, value etc. You do not want to waste time searching for the right component!

6) Use a "common" or "ground" connection for your entire circuit, its power supplies, and its instruments (for which this is normally the case!). Ground is normally the reference against which most measurements will be made. Thus, you want ground connections to be a unquestionably solid and reliable.

7) Use "bypass capacitors" on your prototyping board. These capacitors connect between the power-supply line and the "common" or "ground" lead. A parallel combination of a large polarized capacitor (say 100 μF, Tantalum) and a small non-inductive capacitor (say 1.0 or 0.1 μF, Ceramic) for each supply is a good idea.[6] You want the power supply to be ac ground at both low and high frequencies.

8) Use 10× probes with your oscilloscope (rather than wires) for convenience, flexibility and reliability. Note that many circuits misbehave with long wires connected to their high-impedance nodes. A 10× probe reduces the loading on your circuit.[7]

9) Use external triggering of your oscilloscope, through a direct connection from a fixed-amplitude output on the signal generator. This allows you to maintain display integrity while moving your probe from node to node, while modifying the circuit, while changing signal

---

[5] Experiment #0 is intended to help you with this familiarization,

[6] Bear in mind that the wire connections to the power supplies are inductive, particularly for high-frequency load currents.

[7] If you need more gain to allow you to see very small signals, you can consider a 1× probe, but *beware* of the possibility that you will change circuit behaviour.

10) Generally speaking, during experimentation or circuit troubleshooting, make only *one change at a time*, cheking what happens between each change. Otherwise you may miss the experience with cause-effect relationships that **Laboratory Exploration** is all about.

Good Luck! Bonne Chance! ¡Suerte!

VI ACKNOWLEDGEMENTS

As may be obvious to you, this **Manual** is the result of a great deal of work by many people. I wish to extend my specific thanks to some of them:

- To all of those contributed to the first edition.
- To Laura Fujino, the love of my life, for countless contributions in document preparation, correction, organization, production, management, as well as to state of mind, and general well-being;
- To Franky Leung, who helped with the vagaries of workstations, computer networks and UNIX; and figure preparation.
- To Raymundo Tang Tang, who has prepared the "Electronics Workbench" input files, and done many simulations.
- To Bill Zobrist, my editor, who introduced us to "Electronics Workbench".

To these and to others, who contributed in other ways, and to the countless students in the past who provided the motivation for my interest in laboratory instruction, and have endured many of the Experiments, I am indebted. As well, I am grateful to the Department of Electrical and Electronic Engineering, Hong Kong University of Science and Technology, whose facilities and services have been used so intensively in this work. Finally, I am indebted to Professor Adel S. Sedra who has always motivated me to do more of what I should, and less of what I would!

But, for the errors and omissions that you will doubtless find, I, alone, am responsible. For them, I must apologize, and thank you in advance for your tolerance and goodwill in enduring and reporting them. For this **Manual**, as the process it describes, must be a living and growing thing, in whose upbringing, improvement, and discipline, I welcome your continued interest.

Kenneth Carless Smith
Department of Electrical and Computer Engineering
10 King's College Road
University of Toronto
Toronto, Ontario, Canada
M5S 1A4

(416) 971-2286 (FAX)

lfujino@cs.toronto.edu   http://www.sedrasmith.org

August 1997

EXPERIMENT #0

GETTING STARTED – INSTRUMENTS AND MEASUREMENT

I OBJECTIVES

The intent of Experiment #0 is twofold: First it is is to introduce you to the generalities of the instruments you will encounter in your laboratory. Second, it will be to practice with some of their more basic features in the context of measurements made on simple passive circuits. Where necessary, reference will be made to other sources of information: These will include general Appendices to this Manual, at one extreme, and the complete instruction books of the actual instruments available in your laboratory, at the other. As well, your instructor may have prepared summary sheets for individual instruments to which you have access. Finally, you are encouraged to make use of the software product called "Electronics Workbench",[1] in which simulated instruments are an important visualization tool.

II COMPONENTS AND INSTRUMENTATION

A basic laboratory instrument set should include[2]

a) 1 – Prototyping Board (PB).

b) 2 – Power Supplies (0-20 V, 0-100 mA at least).

c) 1 – Digital Multimeter (DMM).

d) 1 – Dual-Channel Oscilloscope ($\geq$ 50 MHz).

e) 1 – Function Generator ($\geq$ 1 MHz).

As well, a second DMM, a pulse generator, and a frequency counter would be quite useful. Finally, a characteristic-curve tracer and a waveform analyzer would be often informative, on a shared-time basis.

For reference, Fig. 0.1 shows the front-face displays of the instruments provided in "Electronics Workbench". The functions of the curser-selectable simulated control areas you see there should become obvious as we proceed here. Otherwise, consult the software package[3] for detail.

Components[4] for this particular laboratory are as follows:

a) Resistors:

- Two each of 10 Ω, 100 Ω, 1 kΩ, 10 kΩ, 100 kΩ, 1 MΩ, 10 MΩ with ¼ watt rating [it would be ideal if the 1 kΩ and 1 MΩ have 1% tolerance].

- Two of 1 kΩ with 1 W rating.

- One each of a sampling of resistors in both the 1% and 5% series with values between 10 kΩ and 100 kΩ.

---

[1] "Electronics Workbench", by Interactive Image Technologies, Ltd., Toronto, Ontario, Canada, is an interactive-laboratory simulation tool, running under Windows 95.

[2] For example, typical commercial products are: a) Global PB-103, b) Tektronix PS280, c) Tektronix CDM250, d) Tektronix 2205, e) Tektronix CFG250 and a partial combination of a), b), e): Global PB-503. Of course, there are many other alternatives.

[3] A student sampler for this programme is contained in the CD-ROM packaged with the Fourth Edition of "Microelectronic Circuits" by Sedra and Smith, Oxford 1997.

[4] While component tolerance is not a very critical issue here in Experiment #0, 1% resistors are recommended in general as economical aids to understanding. Precise capacitors are likewise useful, but possibly too expensive. Premeasured and selected components are an alternative. Use of a digital ohmmeter (and possibly a digital capacitance meter) is recommended both before and during the experimentation process.

b) Capacitors:
- One 100 pF.

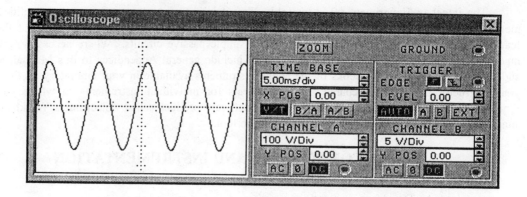

 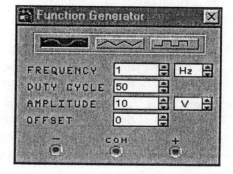

Figure 0.1: Front-Panel Appearance of Instruments Provided in "Electronics Workbench"
a) Oscilloscope; b) Digital Multimeter; c) Function Generator

III READING

General familiarity with your Text and its Appendices, particularly Appendix F, and with this Manual, overall, would be useful. In this Manual, read the Preface, this Experiment, and the Appendices. As well, make yourself familiar with "Electronics Workbench" if that software is available to you.[5] Finally, note that in this Experiment like many to follow, the most important thing that you can read is the experiment itself, particularly the Explorations!

IV PREPARATION

Following the usual pattern in this Manual, **Preparation** tasks are keyed directly to the **Explorations** to follow, using the same section numbering and titling but with a P prefix.

As noted above, the most important Preparation you can do in this and other experiments is to actually read the experiment, particularly the Exploration part, very early in your preparation process. Do so relatively

---
[5] See Footnote 3.

quickly at first, to see what is there, just skimming the measurements to get an idea of where they are headed. Then go back for more thorough reading as time permits. In this Experiment, a somewhat special attempt to provide background on which to build, much of what is included is for reference in later Experiments when you have to face up to the need for good experimental practice. Now, try to read and understand but recognize that much of what is said is quite abstract until you actually face the problems discussed. For this reason the general discussion is interlaced with sections called **Measurement** which are intended to provide a more active learning experience.

- ## THE TOOLS FOR THE TASK
P1.1 The Digital Multimeter (DMM)

(a) A particular digital ohmmeter uses a DVM circuit and a constant-current supply to measure resistance. Sketch the circuit arrangement arranged to measure resistor R.

(b) If the most-sensitive range on the DVM is 1.99 mV full scale, what current do you need to create an ohmmeter with a 1.99 kΩ scale? What constant current would you use to create an ohmmeter with a full scale reading of 19.9 MΩ, for which the voltage across an open circuit is limited to to 2.5 V?

P1.2 The Prototyping Board (PB)

(a) Read any description you may be given about the details of your prototyping board.

(b) How many isolated five-socket strips do you need on your PB to connect ten resistors, totally isolated from one another. If the resistors were all connected in parallel, what is the minimum number of five-socket strips that you need (be careful!). How many strips do you need if they are all connected in series? What if they are connected as a series of two groups of five resistors in parallel?

P1.3 The Power Supplies

(a) You have available two isolated power supplies capable of provided outputs from 0 V to 20 V at up to 100 mA. Representing each of these supplies symbolically by a rectangular box with two terminals, one marked (+) and one marked (−), and a number indicating the voltage, provide sketches to show the following supply systems, one node of which is connected to ground.

  i) A ± 15 V supply pair.

  ii) A single + 30 V supply [This solution is not unique! Why?]

  iii) A single ± 10 V supply with voltage variable over the whole range from − 10 V to + 10 V using *a single control.*

P1.4 The Oscilloscope

(a) A particular circuit node has an average voltage of 100 V with a signal component consisting of a somewhat randomly-occuring 0.1 V 20 ms positive pulse with a maximum pulse-to-pulse interval of 200 ms. Provide a list of various oscilloscope control settings which produce the following displays on the 10 × 20 unit oscilloscope screen:

i) A positive pulse five units high with its lower end on the second screen division from the screen bottom, and two units wide, rising at the second screen division from the left.

ii) As above, but showing in greatest possible detail the 2 ms period immediately after the pulse has fallen, with node-voltage reductions directed downward on the screen.

iii) The largest-possible view of the node voltage including its dc value, where 0 V is adjusted to be at the second screen division from the bottom, with the display guaranteed to include at least two (very small) pulses.

P1.5 Oscilloscope Application Notes

(a) An oscilloscope channel input has an equivalent circuit represented by a parallel combination of a 1 MΩ resistor and a 20 pF capacitor. Your challenge is to design the two components of a 10× probe. [Hint: These consist of a resistor and a capacitor in parallel.] Sketch the circuit. What are the component values needed? What is the equivalent circuit as seen at the probe tip?

P1.6 Function Generator

(a) What are the maximum rates of change in a 10 Vpp signal at 100 kHz, if the waveform is a) sinusoidal, b) triangular, c) square, as produced by a system having a 20 MHz bandwidth?

- **MORE-GENERAL FAMILIARIZATION EXPERIMENTS**

P2.1 The Oscilloscope with the Function Generator

(a) What bandwidth does an instrument output stage need to provide a square wave having 50 ns transition times, assuming that the internal circuits are quite ideal.

(b) A designer wants to provide a 1 mA constant current to a very-low-resistance load for 1 ms periods at 1 ms intervals. She decides to use a function generator and 0.1 µF capacitor. What is her solution likely to be?

(c) It can be shown that the displayed rise time of an oscilloscope is the square root of the sum of the squares of the rise times of the observed signal and of the oscilloscope itself. For what rise time of the square-wave output of a function generator is the displayed value correct within 10% when displayed on a 100 MHz oscilloscope? [Hint: See Eq. F.13 in Appendix F of the Text.] What is the fastest displayed rise time on a 150 MHz oscilloscope?

P2.2 Secondary Properties of the DMM and Oscilloscope

(a) Consider the circuit of Fig. 0.2, to which a 10 V dc source is connected. What voltage is measured by: i) A DVM with a 10 MΩ input resistance; ii) An oscilloscope having a ×1 probe with a 1 MΩ input resistance.

(b) Repeat (a) for a 10 V peak 100 kHz sinewave for the situation in which the capacitances of the DVM and the ×1 probe are 10 pF and 80 pF respectively.

V EXPLORATIONS

- **THE TOOLS FOR THE TASK**

E1.1 The Digital Multimeter (DMM)

- **Background:**

Your Digital Mutlimeter (DMM) should be a portable, battery-operated multiple-range instrument including at least voltage and resistance capability. It usually also includes current measurement, and ac voltage (and current) measurement, as well. It sometimes includes capacitance measurement.

- **Measurement:**

a) Examine your DMM. Be prepared to answer the following questions: How many digits does it have? {The leading 0/1 is referred to as a half digit.} How many different major functions (including voltage, current, resistance, etc) does it perform? What kind of ac measurement does it make (eg, true rms, or rms-calibrated peak)? How many voltage/current/resistance ranges does it have? What are their full-scale values?

Experiment #0–5

b) Begin to use your DMM as an ohmmeter (a DOM!). Measure a wide range of resistor values (eg 10 Ω, 1000 Ω, 100 kΩ, 10 M) on each available range, but particularly the one giving the greatest precision.[6] Note the variation of your measured resistors from their nominal value. Try a second resistor of the same value to get a sense of unit-to-unit device variability. Try other values of resistors to practice your use of colour codes.

c) Later, you will use your DMM for its most important role, "the measurement of voltage" where it is usually referred to as a DVM.

d) Finally, notice that your DMM is (normally) battery-operated, and can be connected anywhere to almost any circuit. But, BE AWARE of SAFETY![7]

e) With a second DVM, (perhaps borrowed from a neighbouring experimenter), measure the voltages across a 1 kΩ resistor which you have connected to your ohmmeter. Try different ohmmeter ranges. Try a 1 MΩ resistor. Notice the polarity of the voltage your ohmmeter produces.[8]

E1.2 The Prototyping Board (PB)

The preferred type of prototyping board consists of a number of white plastic molded parts with an array of funnel-topped holes leading to a set of metallic strip-spring interconnects inside, making a reliable socket for component-lead-size wire. Various interconnect patterns are available. One pattern consists of parallel columns of a linear cluster of five interconnected sockets oriented at right angles to a trench across which dual-in-line IC packages can be placed. As well, typically at the edge of the board, running in parallel with the central trench, there are longer strings of interconnected sockets which can be used as power-supply buses.

The preferred scheme is to have a number of such module boards mounted on a metal plate (which can provide a degree of electrostatic shielding to the assembled circuits), and equipped with binding posts and coaxial connectors for convenient connection to power supplies and signal generators. It is ideal also if the PB also includes one or more multi-pole switches for control of power supplies, since it is good practice *never* to make significant changes to a circuit with the power connected.

A word of warning is in order: It is very easy to damage the sockets of your board by inserting wires that are too large. Such wires occur on power resistors for example. Resistors rated at 1/2 W and 1/4 W are ideal. Those rated at 2 W have leads which are too big – **avoid them!**

• Measurement:
   a) Now use your ohmmeter with two small pieces of wire connected as probes, to identify the connection patterns on the boards you have available. As well, explore the connectivity of any auxillary terminals and switches your prototyping system may include.

O **Prototyping Board (PB) Wiring**

There are many different approaches to PB wiring. It is possible for example to use very short leads dressed to the surface of the board that are thereby very neat and secure. But this style is not appropriate for most experimentation, where there is usually a desire to reuse components and wires, and accordingly not to shorten their leads (at least very much). However, the arrangement you use should be neat and orderly. Otherwise, it is very hard to troubleshoot or modify.

Generally speaking, a good PB layout should follow the "natural shape" of the circuit drawn in the Sedra and Smith style. This necessitates a positive bus at top, a negative one at the bottom and a ground one in the

---

[6] Note that the auto-ranging function on some meters automatically does this optimization.

[7] Though your laboratory is generally quite safe, voltages inside the power system around you and inside the cases of some of your instruments are *very dangerous*!

[8] Note that for a voltmeter, the red terminal is the positive one. However, this is often not the case for an ohmmeter. That is why you are checking!

middle. Since the latter is not possible with usual PB designs, ground at the bottom *and* at the top is a good idea, since ground is the most important common element in a circuit, and should logically be more substantial than the other power buses in good wiring practice. However, ICs make some aspects of good wiring practice very difficult to follow, and the resulting layouts very messy unless special steps are taken:

Such steps include using relatively low-complexity IC packages, one op amp per IC for example. Another approach is to run some wires from the actual pin to create a "surrogate node" located somewhat more logically in the context of the ideal conventionally-drawn circuit schematic. This idea of "surrogate nodes" can also be extended for use in conveniently locating components that must be changed frequently. Such nodes are also a good idea, for use in conjunction with an insulated wire, for components that would otherwise span a large part of the board and be in danger of short-circuiting to the bare leads of others.

E1.3 The Power Supplies

For most Experiments in this Manual, you need two power supplies. They may be separate or packaged together as a dual supply (or even a triple supply). The outputs of each supply should be "isolated", that is not connected together or to anything else. But note that in some dual-and-triple supply units, all supplies share a *common* connection. While this is less flexible in general, it is acceptable for all of the Experiments in this Manual.

Each supply should have a voltage control, although often there is both a "fine" and a "course" control. For the purposes of this Manual, the voltage range should include 0 V and 20 V. The supply should also be current-limited, so that accidental short-circuits can be tolerated. For this purpose, the most useful supplies provide a current-limit control. If available, it is good practice to adjust it at an early stage of experimentation to a point which allows the supply to operate normally in voltage mode except when something unexpected occurs. In that event, a need for current beyond the set limit causes the output voltage to fall, while the current is held constant (or even reduces!):

- **Measurement:**

    a) Examine one of your supplies. Note the two (isolated) output terminals [a positive (often red coloured) one, and a negative (black) one]. You will also see another terminal, usually called "ground", which is *not* connected to the other two, but to the chassis, or the metal frame or case of the supply. Use your ohmmeter (on a very high-resistance range) to check this.

    b) Wire the ± output of your power supply (PS) to your prototyping board (PB) to make subsequent measurement convenient. Convert your DMM to a voltmeter (a DVM) (remember you have just used it as an ohmmeter, or DOM), for this step. Measure the voltage at the output of the power supply as connected on your PB, while the supply is turned on and you are adjusting the voltage control. Connect a 1 kΩ 1 W resistor across the supply connection on your PB, and repeat the voltage measurement. Note that the current-limit control must be adjusted upward to get the full voltage range. Now, with the voltage control adjusted to provide + 10 V to your resistor, lower the current-limit control to the point where 10 V begins to droop, and then raise the limit slightly. Now, shunt your 1 kΩ resistor momentarily with a second one and note the drop in voltage. To what value? What is the current now flowing in the two parallel 1 kΩ resistors?

    c) Leaving the controls of the supply in the position just established in step b), remove the resistors, but not the DVM. Now, short the supply terminals! What happens? Now, **disconnect** your DMM and switch it to measure current (as a DCM), on a relatively large range. Now use it to short-circuit the supply! What is the short-circuit current you find?

    d) NOTE, SEEK YOUR INSTRUCTOR'S GUIDANCE AND PERMISSION FOR THIS STEP. Remove the meter connections from the supply terminals on your PB. Set the voltage and current controls very near, but not quite at, their lower limit. Arrange your DMM as an ammeter on its *largest* scale. *Short-circuit the power supply with your ammeter!* Adjust the supply current limit carefully upward to check its available range. Be very careful not to exceed the ammeter's full-scale reading. At the largest possible current reading (limited by either your meter or your supply), vary the supply voltage control to verify that it has no effect.

E1.4 The Oscilloscope

The oscilloscope provides a two-dimensional display whose axes are referred to as vertical and horizontal. The display-screen activity is best understood in terms of a moving spot which paints on the surface of the display. The vertical motion of the spot is moderated by the so-called vertical-channel controls of which there are usually two sets called channel A and channel B. Each set includes an input connector, a selector switch, a polarity switch, a calibrated attenuator control, a continuously-adjustable-gain control, and a vertical-position control. A shared selector switch can select one channel, or the other, or their sum, or both, to appear on the screen. As well, there is a choice referred to as chopped/alternate which controls the way the two channels share one display spot. In "chopped mode", each channel has control of the spot for a short interval, after which the other channel is given its turn. The timing of this exchange is relatively random. In "alternate mode", the interval for which each channel controls the spot is coordinated with the time base (to be discussed shortly) to allow the display to represent the activity of a single channel for a suitably longer period. Each mode has advantages. Generally speaking, chopped is good for slowly-evolving signals, while *alternate* is good for rapidly-changing ones.

The horizontal motion of the spot is moderated by the horizontal channel controls. These controls allow the controlling signal to be one of four choices: a built-in "time base" in which a repetitive linear-rising ramp signal is created, either vertical channel, or an external signal. The latter two choices allow the display to be used, for example to plot A versus B, while the former, provides the conventional scan display consisting of a plot of A or B or both versus time T. The time base function is typically controlled by a selector switch for control of major sweep-rate steps, in conjunction with an interpolating fine control. As well, there is a horizontal-position control. Overall, the time-base operates under control of an initiating signal called a trigger.

A trigger-source selector allows this to be either channel A, channel B, the power line, or an external signal. A trigger-slope switch allows the rising or falling edge of a signal to be selected. A level control selects the ± voltage level of the selected signal at which triggering is to occur. As well, there are various kinds of automatic triggering systems which remove the need for much manual adjustment.

E1.5 Oscilloscope Application Notes
A. Signal Connectors and Ground

Note that the connectors at the channel inputs are coaxial (using so-called BNC connector, as do the trigger and horizontal inputs, as well). For all of these, the outer (reference) connection is to the chassis (or case) of the oscilloscope, and the inner connection is the relatively high-impedance input to the channel itself. Thus the oscilloscope is quite unlike, for example, a battery-powered DMM whose connections are quite symmetric, neither one being more bulky than the other. Correspondingly, the oscilloscope's connection to a circuit is *always* asymmetric, as well: Thus the scope ground (often connected as well to the power-line ground!) is always connected to (or possibly defines) the test-circuit ground. Only the inner conductor of the channel input has a low enough capacitance to be connected to a sensitive circuit node. But even that is not ideal! {See the discussion of probes following.}

B. Use of Oscilloscope Probes

Oscilloscope probes serve several important purposes: First, since they are implemented with shielded wire, they tend to limit the electromagnetic interference (EMI) that a single-wire lead would bring to the input. Second, instrument ground can be accessed near the probe tip allowing a signal and its local ground reference to be both connected to the oscilloscope, eliminating the loop formed by two separate wires and the EMI for which such a connection is a loop antenna. Finally, within the probe body, a series-connected parallel-RC network forms a special frequency-compensated voltage divider (one called a compensated attenuator) which raises the input impedance at the probe tip, thereby reducing circuit loading. Usually a capacitor in the probe can be adjusted to make the probe attenuation frequency-independent.

There are several types of probes available, including ones referred to as ×1, ×10, ×100, as well as paired combinations with switch selection. Of these the ×1 − ×10 combination is probably the most useful. Otherwise the ×10 is best. Unfortunately, the ×1 probe lacks one of the three benefits that probes can bring. Rather than raising the input impedance by reducing the input capacitance, a ×1 probe actually *increases* the capacitance greatly, simply because of the shielded cable it uses. Thus the ×10 probe, whose capacitance is only slightly more than 1/10 of the regular channel input capacitance, is a very good choice. However, it does bring one disadvantage, and that is a factor of 10 *signal loss*. Unfortunately, while the probe is labelled ×10, there is no amplifier (usually) in it. Rather, there is a resistor-capacitor network with a 20dB *loss*. Ironically, the ×10 refers to the need to multiply the gain settings of the oscilloscope channel attenuator by 10, to become, for example 50 mV/unit of screen display, rather than the 5 mV/unit that can be obtained without a 10× probe. Thus, in general, use ×10 probes to reduce circuit loading, but only if the loss of a factor of 10 in signal amplitude can be tolerated!

C. Channel Gain Calibration

It is important when you use your oscilloscope to have confidence in its calibration. For assisting with that confidence, many oscilloscopes provide a calibration terminal which allows you to verify the channel gain, and also to adjust the frequency compensation of the input probes. Lacking that, it is possible to calibrate using a a dc power supply and a DVM.

However, for many purposes, in making comparative measurements with two probes, it is essential that both channels have *identical calibration* over the full frequnecy range, not necessarily perfect, just *identical*. Thus a good general idea during measurement is to place *both probes* on the same circuit node to verify that they both convey the same truth! If they do not, there are some things to be done: i) The problem may lie in the probe compensation adjustment. But do not adjust the probes unless the signal you are measuring is appropriate − a good square wave is best! ii) You can use the continuous-gain control available on each channel, which normally rests at its maximum (calibrated) position, to equalize the gain in the two channels, and iii) You can report the instrument to the laboratory instructor as needing repair!

D. Channel "Normalization"

"Normalization" is a formal name for the process described loosely in the contest of channel gain checking and equalization in c) above. It is very very useful in comparing two signals that are somehow related, but not simply identical, and allows visual differencing, a process of importance, for example, in examining amplifier distortion. In this important amplifier application, the normalization idea is generalized somewhat by first equalizing two displayed signals at the input and output of an amplifier for relatively low signal levels where distortion is unlikely. Then signal levels are raised to induce distortion, while the step attenuators of *both* channels of the oscilloscope are adjusted to compensate for increasing signal size. Assuming that the channel attenuators are equal, and at least linear, the details of distortion become directly visible as the visual difference between the two displayed signals.

To be concrete and specific, "normalization" consists, in its basic form of the following steps: i) Connect the probes of both channels to the same ac signal. ii) Adjust the channel attenuators so the displayed signals are roughly the same and of satisfactory magnitude. iii) Now adjust the vertical position of the channels and the fine-gain control of the larger signal until both *exactly* overlap.

Normalization can be a very flexible comparison technique when used in conjunction with the channel step attenuator, the channel polarity-reversing swtich, and external voltage dividers for sizing of input test signals.

E. AC and DC Coupling

Each channel of your oscilloscope provides an option for either direct coupling (dc) or ac coupling. In the ac-coupled mode, a very large capacitor is inserted in the channel internally in series with the input connector. In the ac-coupled mode, the dc value of the input does not affect the display. Usually ac-coupling is used to examine small signals on a large dc base. If the channel is direct-coupled, a gain setting high enough to allow fine detail to be seen, can move the signal off the screen beyond the ability of the vertical position control to recover it. Though ac coupling is essential in such a case, it is generally *best to use direct coupling*! This allows you to keep track of operating conditions better. Often, using the vertical position controls is enough to solve the problem.

F. Taking Difference Measurements

Differences are very important in electronic measurement. "Normalization" discussed in D. above can be seen (!) as a visual differencing technique. But so also is ac-coupling a difference technique! In ac coupling, one has simply subtracted the dc value of the measured voltage [which is stored on the capacitor] from the total signal to emphasize the signal part! But there are other differencing schemes: The most elegant is the differential input with which some oscilloscopes are equipped. This uses an electronic differencing technique that we will learn more about later in Experiment #2 where we study operational difference amplifiers. But such an oscilloscope is relatively rare and expensive, and does not actually perform very well in other ways, for various reasons. However most oscilloscopes have a poor-man's approximation to differencing. This is made available by the ability of many oscilloscopes to display $(A + B)$ and to invert A or B or both to obtain $(A - B)$ [or $(B - A)$]. While this difference is far from perfect and with a very limited high-frequency response, it is a useful alternative to ac coupling. For example, within the dynamic range of the channel amplifier, a very samll signal with a high dc average, connected to channel A, can be viewed by connecting a dc power supply to channel B in one of several different ways, and selecting the $(A + B)$ display feature.

E1.6 The Function Generator

A function generator, also called a waveform generator, is typically based on a circuit that you will investigate in Experiment 12 to follow. This circuit naturally produces square, triangular, and sinusoidal waveforms at the same time, with one set of frequency controls. Though all three waveforms are potentially available simultaneously from such an oscillator, the cost of providing output drivers, controls, and connectors normally means that one of the three waveforms is switched to an output circuit incorporating amplitude and offset controls. Most such function generators also provide a fixed-amplitude digital output with TTL/CMOS logic-level compatibility and relatively short transition times. This is a good place to connect the oscilloscope external trigger when using a function generator in testing. Incidentally, if a coaxial cable is used in the connection, the case grounds of the two instruments are automatically well-connected! Note in passing, as implied above, that outputs of a function generator are ground-referenced. However two concessions are sometimes made: One is that an offset control is often provided, which allows the average value of the output waveform (which is normally symmetric around ground) to be made non-zero. The other is that some generators provide a second output which is the 180° complement of the first.[9]

E2.0 MORE-GENERAL FAMILIARIZATION EXPERIMENTS

Our goal now is to perform some larger-scale experiments which illustrate the operation of combinations of the instruments you have available, and to familiarize you with some basic measurement techniques.

E2.1 The Oscilloscope with the Function Generator

- **Goal:**

    To explore the use of two basic tools for signal analysis.

---

[9] Note that such a feature, if available, allows a very convenient demonstration of full-wave rectification!

Experiment #0–10

- **Setup:**
  - Connect the external trigger input of your oscilloscope to the logic output of the generator. Set the oscilloscope to positive-edge external automatic triggering. Wire the output of the generator to your prototyping board. Use the board to facilitate connecting a 1 kΩ resistor across the generator output, noting which end of the resistor is automatically gronded, that is at the potential of the two interconnected instrument cases joined by the outer sheath of the external-trigger coaxial cable. Use a 10× probe on each channel of your oscilloscope.

- **Measurement:**
  a) Set the generator to provide a 10 kHz square wave of 2 Vpp amplitude. Connect probe A to each end of the 1 k resistor to verify which end is grounded. Is it connected to what you will use as a ground bus across your PB? It should be!

  b) Connect both probes to the active end of the 1 kΩ resistor, and adjust each channel attenuator so that each wave covers about half the screen. {Note that the variable controls should be in the calibrated position.} Adjust the sweep speed to display two cycles. What are the settings of the input attenuators and of the time base that you are using?

  c) Adjust the vertical-position controls so that the two waveforms exactly overlap in the centre of the screen. You can use the amplitude control on the generator to align the displayed waveforms with the screen's calibration markings, while maintaining amplitude somewhere near 2 Vpp. Do the waveforms look ideally square? If they overshoot or undershoot immediately after the transition, your probes need adjustment. Seek instruction on how to do this!

  d) Do the two traces exactly overlap? Is one bigger than the other at the end of each half cycle? By how much? If the difference is great, check that something may be wrong: wrong probes, wrong settings, the fine control not at the calibrated position, etc. If you don't find the problem, ask for help; Something is wrong with your oscilloscope or setup. If the differences are not great, lower the gain control on the channel with the bigger output to equalize the displays. You have now succeeded in "normalizing" your display.

  e) Reverse channel B. What do you see? While carefully examining the left-most edge of the screen, increase the sweep speed. Do you see the rising and falling edges? Adjust the triggering level between its new limits and see what happens. Leave the level control where full transitions are visible. Increase the sweep speed, probably as far as possible, to measure transition times defined between 10% and 90% levels of the signal. Do you know the bandwidth of your oscilloscope? If so, use Eq. F.13 in Appendix F of the Text, to evaluate the corresponding rise time. Which has a beter rise time, the scope or the generator?

  f) Now touch the active end of the 1 kΩ resistor with your finger. Does anything happen? Shunt the 1 kΩ resistor with a 100 pF capacitor. Measure the rise time. To what time constant does that correspond (see Eq. F.12)? To what resistor? Where is that resistor? Remove the capacitor.

  g) With the time base set to display two cycles of the input, switch your generator to provide a triangle wave. Try to use the triggering switches and controls to allow you to look at the peaks of the triangle wave. [Hint: Use internal triggering on channel A.] If you succeed, try to examine the top of the triangle. Characterize its roundedness, its symmetry, etc.

  h) With the time base adjusted to show two cycles, reverse the polarity switch on channel A for interest. Leave these switches so that the display looks like a pair of spectacles (spectacular, you might say!). Now try changing the display between chopped and alternate to see if you detect a difference. Now try displaying $(A + B)$. What do you see? What is your interpretation?

i) Repeat some parts of step h) above with sinewaves.

E2.2 Secondary Properties of the DMM and Oscilloscope

• **Goal:**

To explore various secondary properties of the DMM and the oscilloscopoe, including input-impedance, waveform sensitivity and bandwidth.

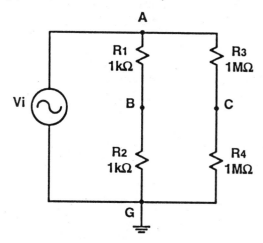

Figure 0.2 An Impedance-Level Test Circuit

• **Setup:**

○ Assemble the circuit in Fig. 0.2 on your prototyping board, with the function generator connected as shown. As well, connect the logic output of the generator to the external trigger connector of the oscilloscope via a coaxial cable.

Measurement:

a) Using 10× probes, display node A on channel A and node B on channel B. Adjust the generator to provide a 2 Vpp sinewave at 100 Hz on node A.

b) Connect your DMM as a voltmeter (a DVM) to ground node G and node B, using a series 10 kΩ resistor as an isolating probe on the DVM lead connected to node B. Note the reading of the DVM. Verify that it is $1/\sqrt{2}$ times the peak value of the signal seen on channel B.

c) Change the generator waveform to a triangle wave and then to a square wave. How do the DVM readings relate to the peak signal at B?

d) Return the generator to a sinewave form and raise its frequency until the DVM reading reaches 0.707 of its 100 Hz value at frequency f_1. As well, for interest, check the DVM reading at $10f_1$ and $100f_1$. What do you conclude about the frequency cutoff of the DVM? {You might also like to check the ac response of your DVM at frequencies below 100 Hz, if you have lots of time!}

e) Extend the idea of step d) in an attempt to measure the bandwidth of the combination of your circuit and the oscilloscope. Remove the DVM to make the next measurement easier to interpret. Raise the frequency of the generator as high as possible in an attempt to find a frequency cutoff, while observing the peak values of the signals at nodes A and B. Refer to the frequency at which something significant happens, as f_2.

- **Analysis:**

    What do you conclude about the upper 3dB frequency of the DVM ac range, and of the oscilloscope when connected to a low-impedance source?

- **Measurement:**

    f) Connect the DVM and probe B to node C. Be sure to use a 10 kΩ "probe resistor", between node C and the DVM lead. With a 100-Hz 2-Vpp sine wave on node A, measure the voltage at node C with your DVM and with channel B. Repeat the measurements with the DVM alone and with the scope probe alone. What do you conclude?

    g) With the DVM connected alone, raise the frequency from 100 Hz to identify its upper 3dB cutoff frequency for this situation, at f_3.

    h) With probe B connected alone to node C, raise the frequency from 100 Hz to identify the cutoff frequency, f_4.

- **Analysis:**

    Since you verified earlier in step e) just above that the oscilloscope cutoff was beyond the range of your generator, what is the cause of this new cutoff? Can it be the probe capacitance? Estimate a corresponding capacitance value.

- **Measurement:**

    i) Repeat step h), with a ×1 probe on node C. At 100 Hz, compare the peak-to-peak values of the waveforms at A and C? What has happened? Raise the input frequency to identify the cutoff frequency, f_4?

- **Analysis:**

    In comparing the results of steps f), h), and g), what do you conclude about the resistance and capacitance presented to the circuit by the ×1 probe? You have now seen two reasons why the ×10 probe is a better choice!

    Before you pack up your equipment, here is a small test of your ability to observe, a very important talent for an efficient experimenter. When raising the frequency in step i), and observing the waveforms at nodes A and C, did you notice anything about the relative timing of the two waveforms? What you might have observed is that the phase of the node-C signal shifted with respect to that at node A. In particular it lagged more and more as the frequency was raised. At what frequency, f_5, was the phase lag 45%? What is the largest phase lag you saw?

I hope this broad introduction to electronics measurement has not been too obscure or overwhelming. If you carry away from it a sense that electronics is non-trivial, but possible, and even interesting, then we are well on the way to better things in the Experiments to follow!

EXPERIMENT #1
OPERATIONAL-AMPLIFIER BASICS and BEYOND

I OBJECTIVES

The primary objective of this experiment is to familiarize you with basic properties and applications of the integrated-circuit operational amplifier, the op amp, one of the most versatile building blocks currently available to electronic-circuit designers. The emphasis will be primarily on the nearly ideal, on what is easily and conveniently done. Exploration of what is less-than-ideal about commercial operational amplifiers will be deferred to Experiment #2, and larger applications to Experiments #8, #11 and #12 where op amps are used as very flexible circuit elements in important electronics subsystems.

II COMPONENTS AND INSTRUMENTATION

Your concentration will be on the 741-type op amp provided, two per IC, in an 8-pin dual-in-line (DIP) package whose schematic connection diagram and packaging are shown in Fig. 1.1.[1] For power, you will use two supplies, +10 V and −10 V, or ±10 V for short. As well, you need a variety of resistors and capacitors, with emphasis on ones simply specified: 1kΩ, 10kΩ, 100kΩ, 1MΩ, 10MΩ and 0.1μF, 0.01μF, 1 nF, and the like. Note that it is important to bypass the two power supplies directly on your prototyping board, using, for each supply, a parallel combination of a 100 μF tantalum or electrolytic capacitors, and a or 0.1 μF low-inductance ceramic capacitor. For measurement, you will use a digital multimeter (DMM) with ohms scales, a two-channel oscilloscope with ×10 probes, and a waveform generator.[2]

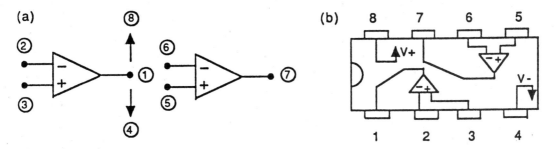

Figure 1.1: The MC 1458 − Two 741-Type Op Amps in an 8-pin DIP
(a) Block schematic
(b) Top view of the dual-in-line package (DIP) with internal connections shown

III READING

Sections 2.1 through 2.6, of the Text, are related directly to this Experiment. While not all issues discussed there are explored here uniformly, broad familiarity with them will allow you to identify areas for concentrated reading as the need arises. The order of coverage here closely follows that in the Text.

---

[1] Device data sheets are available through the Web site: www.sedrasmith.org, as well as in the ancillary Manual "A Pracitcal Guide to Selecting Electronic Components", Oxford University Press, 1997, by Wai-Tung Ng.

[2] See Experiment #0 for general information about instrumentation and measurement, as well as Appendices A and B.

IV PREPARATION

As the name implies, **Preparation** is intended to help familiarize you with the experimental work to follow. Ideally, by raising questions about the specific circuits you will later explore, it will help you in thinking about the experiments you will perform, and the results you will obtain, **as you proceed**. Note the emphasis! An experiment can (and should) be a process of active discovery, one in which thinking and doing are conjoined; in short, a process of "hypothesis and test". Otherwise, treated procedurally, without the mind "in gear", so to speak, blind laboratory measurement is work for slaves, not for the master you wish to become!

For you convenience, the **Preparation** directions will be numbered to correspond to sections of the **Explorations** following. You will note that, in general, quite a lot of **Preparation** work is specified. Sometimes, depending on your other assignments, will not have enough time to do it all. However, it is presented to pique your interest, and to inform you of some aspect of the direction in which the practical exploration will go. Expect the advice of your instructor about what to do in detail. Otherwise, think about the solution of all the **Preparation** questions first; then solve some of the more interesting ones. As well, of course, you can prepare by simulating some of the Explorations with PSPICE, or using "Electronics Workbench"[3]. Again it is expected that your instructor is the one who is best able to advise and direct how much work he/she wants you do to.

Unless otherwise specified, in what follows, assume all op amps to be ideal.

• THE INVERTING AMPLIFIER

P1.1 DC Voltages and Gain

(a) For the inverting amplifier circuit to the right of node B in Fig. 1.2, what is the expected closed-loop gain (as measured from node B to node D)? What is the input resistance to the right of node B?

(b) For the test adapter shown to the left of node B, and employing resistors R_a, R_b, what voltage is produced at node B, for a node A input of +10 V? −10 V? Ignore the loading effect of R_1.

P1.2 Quick Changes of Gain

(a) Design an op-amp circuit with an input resistance of 1 kΩ and a gain of −5 V/V. What are the values of R_1 and R_2 you have chosen?

P1.3 AC Gain and Overload

(a) For the situation described in E1.3, with R_a reduced to 100Ω, calculate the gain from node A to node D.

P1.4 Virtual Ground

(a) Consider the basic inverting op-amp circuit shown at the right of Fig. 1.2, with a 91 mV peak signal applied at node B. For an ideal op amp, what signals would you measure at nodes C, D?

(b) For an op amp with an open-loop gain of 1000 V/V and the same **output** at node D as found in (a), what would the voltage at node C become? In this situation, for what value of resistor shunted from node C to ground does the current in R_2 (and thus the voltage at node D) reduce by 10%?

P1.5 Output Resistance

(a) An op-amp circuit whose output is 1.0 V peak with no load reduces by 15 mV when it is loaded by a 100Ω resistor. Estimate the output resistance of the circuit.

---

[3] To make this easier, the circuits in this Manual are being prepared in "Electronics Workbench" format, for later electronic distribution.

- **THE NON-INVERTING AMPLIFIER**

P2.1 DC Voltages and Gain

(a) Using an ideal op amp, design a non-inverting amplifier with gain of +11 V/V having low currents in the associated resistor network, but with no resistor larger than 10 kΩ.

P2.2 Quick Changes of Gain

(a) What is the gain of the circuit in Fig. 1.4, from node B to node D, with R_1 shunted by a resistor of equal value? With R_2 shunted likewise? With R_2 shorted?

P2.3 AC Gain and Input Resistance

(a) For an *ideal op amp* (having very very high gain) in the unity-gain non-inverting amplifier topology, what is the voltage between the + and − input terminals for normal operation? For a 1 kΩ resistor shunting the ± terminals what input current would flow at node B for ±1 V signals at the output? What is the corresponding input resistance?

(b) Repeat (a) under the condition that the op amp has an open-loop gain of only 100 V/V.

- **A GENERAL-PURPOSE AMPLIFIER TOPOLOGY**

P3.1 Individual Inputs, Difference Gains

(a) Calculate the expected gains for individual inputs A, D, F to output C, of the circuit shown in Fig. 1.4.

P3.2 Common-Monde Gains

(a) For what two inputs of the circuit described in E3.1, are the gains equal in magnitude?

V EXPLORATIONS

- **THE INVERTING AMPLIFIER**

E1.1 DC Voltages and Gain

- **Goal:**

To explore the basics of inverting-amplifier operation, and the occurence of virtual ground.

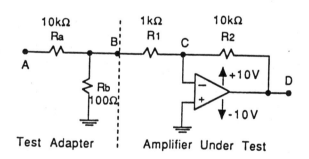

Figure 1.2 A Basic Inverting Amplifier with Input Attenuator (For Testing)

- **Setup:** {*Note that R_a and R_b form a so-called input attenuator, which allows you to provide relatively small signals at the amplifier input (node B) without requiring that the source be able to produce them directly.*}
  - ○ Assemble the circuit as shown in Figure 1.2.
  - ○ Adjust the supplies to ±10 V using your DVM.

- **Measurement:** *[Use your DVM to measure nodes B,C,D in turn.]*
  - a) Node A open (or grounded); Measure B,C,D.
  - b) Node A connected to +10 V; Measure B,C,D.
  - c) Node A connected to −10 V; Measure B,C,D.

- **Tabulation:**
  V_A, V_B, V_C, V_D, for V_A = 0 V, 10 V, − 10 V.

- **Analysis:**
  Consider the location of virtual ground. Calculate two estimates of the voltage gain, v_D/v_B.

E1.2 Quick Changes of Gain

- **Goal:**
  To practice component shunting as a measurement technique, thereby identifying the role of each element in a circuit.

- **Setup:**
  - ○ Use the circuit as shown in Figure 1.2, with node A connected to +10 V.

- **Measurement:** *[Continue to measure node D with your DVM.]*
  - a) Shunt resistor R_2 by one of equal value to reduce the gain by a factor of 2; Measure D, B.
  - b) Shunt resistor R_1 by one of equal value to raise the gain by a factor of 2; Measure D, B.
  - c) Open the connection of R_1 to node B, and add a resistor in series with R_1, of equal value, joined to R_1 at a new node to be called X. Measure nodes D, B, X, C.

- **Tabulation:**
  R_1, R_2, V_D, with V_A, V_B, V_X.

- **Analysis:**
  Consider the technique introduced in c) and associated measurements as a way to verify the input resistance of the basic circuit to the right of node B of the unmodified circuit in Figure 1.2. Calculate the input resistance R_{in} at node B.

E1.3 AC Gain and Overload

- **Goal:**
  To explore both linear and non-linear amplifier operation.

Experiment #1–5

- **Setup:**
  - Use the circuit as shown in Figure 1.2, except with $R_a = 1\text{k}\Omega$ and node A connected to a waveform generator,

- **Measurement:** {*Use your two-channel oscilloscope externally triggered from the generator initially, with one probe on node A and the other on nodes B, C, D in turn.*}

  a) Adjust the waveform at A for 2 Vpeak (at 1 kHz); Measure B, C, D. Note the peak values and relative phase of the signals.

  b) Short-circuit resistor R_a; Measure B, C, D. Note the relationship between the signals at B, C, D, using both probes. Prepare a labelled sketch.

- **Tabulation:**

  R_a, v_{bp}, v_{cp}, v_{dp}, v_{Bp}, v_{Cp}, v_{Dp}, v_{Bp}^+, v_{Bp}^-, v_{Cp}^+, v_{Cp}^-, v_{Dp}^+, v_{Dp}^-.

- **Analysis:**

  Consider the effect of attempting to create signals larger than the amplifier's linear output range. Identify the limiting levels at output and input. Note and explain the changes in voltage at node C, when feedback ceases to operate.

E1.4 Virtual Ground

- **Goal:**

  To estimate the resistance of virtual ground.

- **Setup:**
  - As in E1.3 above.

- **Measurement:**

  a) While displaying nodes B and D on your oscilloscope screen, shunt node C to ground with a resistor R of various values, in turn: 1kΩ, 100Ω, 10Ω. Find a resistor which makes a change of 10% or so.

- **Tabulation:**

  v_B, v_C, R, for various values of R.

- **Analysis:**

  Consider the evidence that for a virtual ground, the connected resistance level is not very important, and, correspondingly, that the input resistance at a virtual ground must be very small. Estimate it from your measurements of the peak voltage changes at node D, and the corresponding resistor value.

Experiment #1-6

E1.5 Output Resistance

- **Goal:**

    To characterize the low output resistance of a feedback amplifier.

- **Setup:**

    ○ Establish a setup as in E1.3 above (with $R_a = 1$ kΩ).

- **Measurement:**

    a) While displaying node D on your screen, adjust the generator to provide an output of 0.1 V peak. Now, load node D to ground with resistors chosen small enough to lower the output by a barely noticeable amount (1% or so). Expand the channel vertical scale to make this peak-change measurement more convenient. Estimate the output voltage change with load.

    b) Use your DVM to measure the load-resistor value.

- **Tabulation:**

    v_{d0}, R_L, v_{d1}, δv_{d1}.

- **Analysis:**

    Consider an estimate of the output resistance whose effect you are observing. This output resistance is low because of feedback. You will learn more about this in Experiment #8.

- ## THE NON-INVERTING AMPLIFIER

E2.1 DC Voltages and Gain

- **Goal:**

    To explore the basics of non-inverting op-amp operation, and the behaviour of a virtual short circuit.

- **Setup:**

    ○ Assemble the circuit shown in Figure 1.3. Adjust the supplies to ±10 V using your DVM.

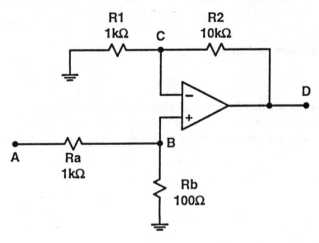

Figure 1.3 A Basic Non-Inverting Amplifier with Input Attenuator (For Testing)

Experiment #1-7

- **Measurement:** *[Use your DVM to measure nodes B, C, D in turn.]*
  - a) Node A open (or grounded); Measure B, C, D.
  - b) Node B connected to + 10 V; Measure B, C, D.
  - c) Node A connected to – 10 V; Measure B, C, D.

- **Tabulation:**

  v_A, v_B, v_C, v_D, for $V_A = 0$ V, 10 V, – 10 V.

- **Analysis:**

  Consider the idea of a virtual short-circuit. Calculate two estimates of voltage gain v_D/v_B.

E2.2 Quick Changes of Gain

- **Goal:**

  To become more familiar with shunting as an exploratory technique in electronics, and thereby extend your understanding of the non-inverting amplifier.

- **Setup:**
  - O As in E2.1, with A connected to +10 V,

- **Measurement:** *[Continue to measure node D with your DVM when making the change; Then measure node B, C.]*
  - a) Shunt R_2 with a resistor of equal value. Measure B, C, D.
  - b) Shunt R_1 with a resistor of equal value. Measure B, C, D.
  - c) Short-circuit R_2. Measure B, D.

- **Tabulation:**

  v_B, v_C, v_D, R_2, R_1, for various combinations of R_1, R_2.

- **Analysis:**

  Consider the gain in each case.

E2.3 AC Gain and Input Resistance

- **Goal:**

  To evaluate the voltage gain and input resistance of a non-inverting op-amp circuit.

- **Setup:**
  - O As in E2.1 with A connected to a sine wave at 1 kHz having 5 V peak amplitude.

- **Measurement:**
  - a) With your oscilloscope, measure the peak amplitude of signals at B, C, D.
  - b) Shunt the op-amp input terminals with a resistor, $R_x = 1\text{k}\Omega$. Measure B, C, D.
  - c) Insert a resistor $R_S = 100\text{k}\Omega$ in series with B and the op-amp +ve input terminal (with the 1 kΩ shunt still in place). Measure B, C, D.

Experiment #1–8

d) Short R_2 and measure again.

• **Tabulation:**

R_2, v_a, v_b, v_c, v_d, R_x, R_S, for $R_2 = 10$ kΩ or 0Ω, $R_x = \infty$ or 1 kΩ, and $R_S = 0$ Ω or 100 kΩ, appropriately.

• **Analysis:**

Consider the gains in each case. Estimate the input resistance (with the 1 kΩ shunt on the input) for $R_2 = 10$kΩ and zero. You will learn more about feedback-amplifier input resistance in Experiment #8.

- **A GENERAL-PURPOSE AMPLIFIER TOPOLOGY**

E3.1 Individual Inputs, Difference Gains

• **Goal:**

To investigate the properties and potential application of a special three-input amplifier which facilitates difference measurements.

• **Setup:**

○ Assemble the circuit shown in Figure 1.4, initially with A, D, F grounded. Adjust the supplies to ± 10 V using your DVM.

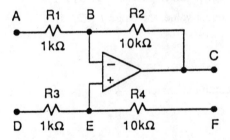

Figure 1.4 *A Multi-Purpose Amplifier Topology*

• **Measurement:**

a) Now, using a sinewave generator to which a 1kΩ - 10Ω voltage divider is connected, generate a 50 mV peak signal at 1kHz.

b) Connect the 50 mV signal in turn to one of A, D, F, separately (the other two remaining grounded), and measure C [and B, E, if you have time].

• **Tabulation:**

v_A, v_B, v_C, v_D, v_E, v_F, for two cases: $v_A = v_D = v_B$ with $v_F = 0$, and $v_A = v_D = v_F = v_{ADF}$.

• **Analysis:**

Consider all three values of voltage gain from the inputs (A, D, F), to output (C), namely v_c/v_a, v_c/v_d, v_c/v_f, and the application of superposition to find the output in terms of the three inputs

Experiment #1–9

acting together.

E3.2 Common-Mode Gain

- **Goal:**

    To illustrate that common-mode signals can be amplified quite differently, and, in one case, hardly at all!

- **Setup:**
    - As in 3.1 above.

- **Measurement:**

    a) Connect the generator to provide a 5 V peak signal to both A, D with F at ground.

    b) Connect the 5 V signal to all of A, D, F. Measure the voltages at C, B, E.

- **Tabulation:**

    v_A, v_B, v_C, v_D, v_E, v_F, for two cases: $v_A = v_D = v_{AB}$ with $v_F = 0$ and $v_A = v_D = v_F = v_{ADF}$.

- **Analysis:**

    Consider the **common-mode** voltage gains from input to output. Find v_C/v_{AB} and v_C/v_{ABF}. To which of these cases does the idea of "common-mode rejection" apply?

Your instructor and I hope that your laboratory experience has been useful and informative, but not onerous.

NOTES

EXPERIMENT #2
OPERATIONAL AMPLIFIER IMPERFECTIONS AND APPLICATIONS

I OBJECTIVES

The objective of this Experiment is twofold: First it will familiarize you with important ways in which the integrated-circuit op-amp departs from the idea! Second, it will allow you to explore, selectively, ways to compensate for some of these deficiencies in applications such as the Miller integrator. You will note that the order of presentation is different from that in the Text in one particular way, a way which is important from an experimental point of view. It is to consider the topics of offsets and offset compensation early, in order that these techniques may be incorporated in later Explorations, where the effects of offsets mightotherwise be troublesome.

II COMPONENTS AND INSTRUMENTATION

As in Experiment #1, our concentration will be on the 741-type op amp provided, two per IC, in a dual-in-line (DIP) package. For convenience, its pin connections are provided in two formats in Fig. 2.1. For power, you will use two supplies, + 10 V and − 10 V, or ± 10 for short. As well, you need a variety of resistors and capacitors with emphasis on ones simply specified: 1 kΩ, 10 kΩ, 100 kΩ, 1 MΩ, and 0.1 μF, 0.01 μF, 1 nF, and the like. Note that it is important to bypass each of the supplies with two capacitors connected in parallel to ground. One of these should be a high-value polarized capacitor, perhaps 100 μF, tantalum; The other should be a relatively large-value low-inductance ceramic capacitor, ideally 1.0 μF, but at least 0.1 μF. For measurement, you will use your digital multimeter (or DMM) as a voltmeter (or DVM) and for its ohms scale, a two-channel oscilloscope with ×10 probes, and a waveform generator.

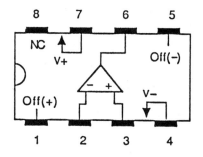 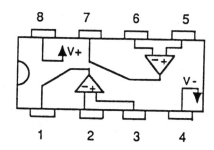

Figure 2.1 Dual Op-Amp Base Connections

III READING

In this Experiment, our concentration will be on Sections 2.4 (particularly 2.4.2), 2.7, 2.8 and 2.9 of the Text. However, as noted above, we will consider the dc problems described in Section 2.9, first, in order that compensation can be considered in later Explorations, as needed. Nevertheless, reading these sections initially in the order presented in the Text, is still a good idea.

IV PREPARATION

In case you have any doubts about the utility of preparation before experimentation, you are advised to read the preamble to **Preparation** provided in Experiment #1 and part of Appendix A on the same topic. As is the case with all Experiments in this manual, this **Preparation** will be keyed directly to the steps in the **Explorations** to follow, with direct reference to the circuit figures and procedures found there.

- **VOLTAGE AND CURRENT OFFSETS**

P1.1 Offset Measurement

(a) A particular op amp has $V_{OS} = 2$ mV, $I_B = 1.5$ μA, and I_{OS} is 200 nA, with reference polarities as defined in Fig. 2.35 and Fig. 2.36 of the Text. It is tested in the circuit shown here in Fig. 2.2, using the resistor values employed in E1.1. Find the values of V_C that are observed, namely V_{Ca}, V_{Cb}, V_{Cc}, following the abc notation of the instruction steps. Note that the polarity of I_{OS} is, in fact, not defined.

- **COMPENSATED MILLER INTEGRATOR**

P2.1 Integrator Offset Control

For the op amp described in P1.1 above, installed in the circuit of Fig. 2.3, find the value of V_D that establishes $V_C = 0$ for $V_A = 0$, assuming that the negative input has the higher inward-directed biasing current.

P2.2 Integrator Operation

(a) The output of the integrator in Fig. 2.3, using a 100 nF capacitor, is observed to move linearly from a rest state of -3 V at time 0, to a final state of $+5$ V, 100 μs later. Describe the input signal that must have been applied.

(b) For the integrator in (a), for what frequency of a sinuosoidal input are the input and output of equal size? At what frequency is the output twice the input? What is the phase relation between input and output in each case?

- **FREQUENCY EFFECTS**

P3.1 Small-Signal Frequency Response

(a) An inverting amplifier with a nominal closed-loop of 100 V/V has a 3dB cutoff at 12.7 kHz. What is it unity gain frequency? What is the 3dB cutoff for a gain of 1000?

P3.2 Slew-Rate Limiting

(a) A particular op amp has a 5-V bandwidth of 100 kHz. What is its slew rate?

(b) For the op amp in (a), what is the highest frequency at which a 2 V peak symmetrical triangular wave can be reproduced?

V EXPLORATIONS

- **VOLTAGE AND CURRENT OFFSETS**

E1.1 Offset Measurement

- Goal:

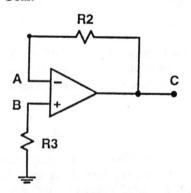

To investigate a simple approach to finding bias current, offset current, and offset voltage by indirect measurement using your DVM.

Figure 2.2 A Circuit for the Measurement of Offsets

- **Setup:**
    - Assemble the circuit in Figure 2.2, using ± 10 V supplies, with $R_2 = R_3 = 1$ MΩ.

- **Measurement:** {*Use your DVM.*}
    - a) Measure V_C for the circuit as wired.
    - b) Short-circuit R_3. Measure V_C.
    - c) With $R_3 = 0$ Ω, add $R_1 = 1$ kΩ to ground from the negative input. Measure V_C.

- **Tabulation:**
    V_C, R_1, R_2, R_3 for three situations.

- **Analysis:**
    Consider the effect of bias current, offset current, and offset voltage on each of these values of V_C, in turn. Estimate each.

- **COMPENSATED MILLER INTEGRATOR**

E2.1 Integrator Offset Control

- **Goal:**
    To explore a practical way to stabilize an integrator circuit, and then to investigate details of integrator operation.

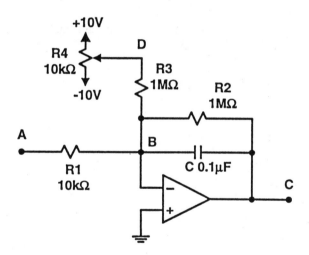

Figure 2.3 A Compensated Integrator

- **Setup:**
    - Assemble the circuit in Fig. 2.3 using ± 10 V supplies with $R_1 = 10$ kΩ, $R_2 = 1$ MΩ, $R_3 = 1$ MΩ, $R_4 = 10$ kΩ and $C = 0.1$ μF.

- **Measurement:** {*Use your DVM for initial dc measurement, primarily of V_C, and then your oscilloscope for signal measurement.*}
    - a) With node A open, and measuring V_C, adjust R_4 to make $V_C = 0$ V.
    - b) Ground node A. Measure node C and node D.
    - c) Measure node C. Adjust R_4 to make $V_C = 0$. Measure node D.

- **Tabulation:**
  Node-A status, V_C, V_D.

- **Analysis:**
  Consider what this compensation process can tell you about the offset voltage and bias current of the op amp. Calculate them. What about the offset current?

E2.2 Integrator Operation
- **Goal:**
  To illustrate the response of the integrator to square and sinusoidal waveforms.

- **Setup:**
  - Use the circuit as shown in Figure 2.3 with compensation adjusted as in E2.1.
  - Connect a function generator to input A.

- **Measurement:** {*Use your dual-channel oscilloscope with external triggering for these measurements.*}

  a) Adjust the generator to provide a 1 kHz symmetric square wave at input A of 1 Vpp amplitude. Measure A and C. Sketch the waveforms, noting peak amplitudes and relative timing.

  b) Switch the generator to provide a 1 Vpp sine wave at input A. Sketch the waveforms, noting the peak amplitude and relative timing.

  c) Adjust the generator to find the frequency at which the signals at nodes A and C have the same amplitude. Note the relative phase.

  d) Find the frequencies at which $|v_c / v_a| = 0.1$ and 10.0. Note the relative phase in each case. You amy want to adjust the input signal level to make the display more convenient while maintaining a sinewave output.

- **Tabulation:**
  Waveform shape, peak amplitude, average value, period, time of each zero crossing as measured from an early rising-edge zero crossing of the input. f_1, Φ_1, f_2, Φ_2, f_3, Φ_3.

Analysis:
  Consider the overall operation of the integrator, particularly the amplitude and phase response for sine-wave inputs. Prepare suitable Bode plots.

- **FREQUENCY EFFECTS**

E3.1 Small-Signal Frequency Response
- **Goal:**
  To explore the small-signal frequency effects in an inverting op-amp circuit.

- **Setup:**
  - Assemble the circuit in Figure 2.4 using ± 10 V supplies. The function generator frequency should be 100 Hz, initially.

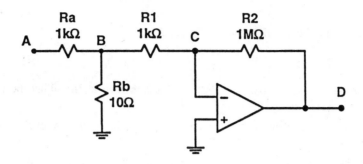

Figure 2.4: A High-Gain Inverting Amplifier for Frequency Measurement

- **Measurement:**

    a) Measure nodes A and D. Adjust the generator amplitude to provide a peak output at node D of 1.0 V at 100 Hz.

    b) Raise the frequency of the generator to the value at which v_D is reduced by 3dB (to $1/\sqrt{2} = 0.707$ of its 100 Hz value). Note the frequency as f_4. [Make sure that the voltage at node A has remained at its initial value as established in a).]

    c) Increase the frequency to $10 f_4$. Measure the peak output voltage.

    d) Change resistor R_2 from 1 MΩ to 100 kΩ and repeat a), b), c).

- **Tabulation:**
    R_2, R_2/R_1, f, v_D for two values of R_2 and six frequencies.

- **Analysis:**
    Consider the relationship between closed-loop gain and 3dB bandwidth of the inverting amplifier. [Hint: Read page 95 of the Text.] What is the upper 3dB frequency of each of the amplifiers tested? What is their Gain-Bandwidth Product? Estimate the unity-gain frequency of the op-amp itself. Sketch a Bode amplitude plot showing both of these amplifiers.

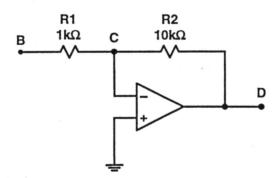

Figure 2.5: A Circuit for Evaluating Slew Rate

E3.2 Slew-Rate Limiting

- **Goal:**
    To explore rate-limited behaviour of an op-amp output for large signals.

Experiment #2-6

- **Setup:**

    ○   Assemble the circuit shown in Figure 2.5.

- **Measurement:**

    a) Measure nodes B and D. For a 1 kHz sinewave input, adjust the input amplitude to provide a 0.1 V peak sinewave at node D.

    b) Raise the frequency to verify that the upper 3dB frequency of this circuit, f_5, is 100 times that in E3.1 b), namely $100 f_4$.

    c) Reduce the frequency to 1 kHz. Raise the input signal until v_D reaches 4 V peak. Note v_B.

    d) Keeping v_B fixed and observing v_D, raise the frequency until v_D falls to 0.707 of its low-frequency value. Note the frequency as f_6; Sketch the waveform.

    e) Lower v_B to half its former value. What does v_D become?

    f) Raise the frequency to reduce v_D to 0.707 of its value in e). Note the frequency as f_7.

- **Tabulation:**

    v_B, v_D, f, for several conditions.

- **Analysis:**

    Consider the effect of slew-rate on large-signal operation. What is the "4-V bandwidth"? Estimate the full-power bandwidth. Estimate the amplitude of the largest signal that can be used with the expected gain at the small-signal upper 3dB frequency.

Op Amps are a bit like people: When they try to do too much, too quickly, they are usually forced to "cut corners"!

EXPERIMENT #3
JUNCTION-DIODE BASICS

I OBJECTIVES

The overall objective of this Experiment is to familiarize you with the basic properties of junction diodes and, as well, provide an overview of some important but simple applications. The main concentration, however, will be on the devices themselves, with most emphasis on their forward-conduction properties.

II COMPONENTS AND INSTRUMENTATION

While many of the Explorations to follow could be done with a single diode of a single type, there is much to be learned about different diode types, and the myriad applications of multiple diodes. Thus you are provided with two IN914, (a small-signal diode), and two IN4004, (a low-power rectifier (diode)). On each, the band indicates the cathode end for normal forward conduction. As well, you have a supply of (standard) resistors, an oscilloscope, a function generator, a multifunction DMM (including ohmmeter ranges), and a dual power supply.

III READING

The Explorations to follow, and the corresponding Preparation are based primarily on Sections 3.4, 3.5 and 3.7 of the Text. Read these relatively thoroughly.

IV PREPARATION

Following the usual pattern in this Manual, the items of **Preparation** are keyed directly to Sections of Part V, **Explorations**, with tasks related to the Figures located there.

- ## DIODE ACTION
P.1.1 Ideal Rectification

(a) For a rectifier circuit consisting of an ideal diode whose cathode is connected to a grounded 1 kΩ load, all fed by a 10 V peak triangle wave at 100 Hz, sketch the input and output waveforms.

(b) Now, augment your sketch with another waveform showing the output with diodes for which the forward drop is constant at 0.7 V, for all currents.

P1.2 Rectififer Filtering

(a) For a circuit resembling that in Figure 3.2, consisting of an ideal rectifier diode, a 100 µF capacitor, and a 10 kΩ load, fed by a 10 V peak triangle wave at 100 Hz, sketch the output waveform.

(b) Augment the sketch in (a) with the waveform corresponding to a diode with a constant drop of 0.7 V.

- ## DIODE CONDUCTION – THE FORWARD DROP
P2.1 Basic Measurements

(a) A particular diode with grounded cathode, fed via a 1 kΩ resistor from a 10 V supply, has a voltage drop of 0.62 V. What is the corresponding diode current?

(b) When the diode operating in the circuit described above is shunted by a 1 kΩ resistor, the diode drop changes by 3.0 mV. What are the coordinates of a second point on the diode characteristic? Estimate n for this diode.

P2.2 Diode Measurement with an Ohmmeter

(a) On a particular digital ohmmeter, having a 1.99 kΩ maximum range, diode (a) reads 14.0 kΩ, while diode (b) reads 12.0 kΩ. Which is the larger junction?

(b) If during the measurements above, the voltages across the diode are measured (using a second DVM) to be 0.70 V and 0.60 V, respectively, what diode currents are flowing?

P2.3 Forward-Conduction Modelling – Finding A Large-Signal Model

(a) Use the information provided by the diode characteristic of Fig. 3.20 of the Text, to estimate V_{DO} and r_D for a diode model which matches operation at 0.5 mA and 5 mA.

P2.4 Forward-Conduction Modelling – Finding A Small-Signal Model

(a) A silicon diode for which $n = 2$ operates at 5 mA with a drop of 0.69 V from a high-resistance source. What is its incremental resistance r_d?

(b) If the junction above is shunted by a 1 kΩ resistor, what new junction voltage would you expect? What shunt resistor would produce a 10 mV junction drop?

V EXPLORATIONS

- **DIODE ACTION**

To set the stage, we will jump directly into an important application of diodes, the one called *rectification*, the conversion of a bipolar (ac) signal into a unipolar (dc) one.

E1.1 Ideal Rectification

- **Goal:**

    To explore the detailed behaviour of the diode in performing the rectifier function.

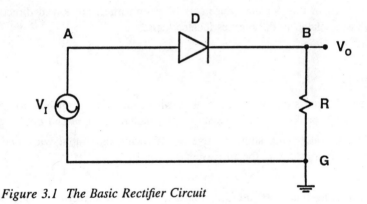

Figure 3.1 The Basic Rectifier Circuit

- **Setup:**

    O Assemble the circuit shown in Fig. 3.1, using a IN4004 diode and 1 kΩ resistor.

    O Set the generator to provide a sine wave at 100 Hz with 10 V peak amplitude.

    O Use your "normalized"[1] two-channel oscilloscope, externally triggered to display waveforms at nodes A and B.

- **Measurement:**

    a) Measure the voltages at nodes A and B, using a "normalized" oscilloscope.

---

[1] See Experiment #0, and Appendix B3.

b) Estimate the diode voltage drop at the peak of the output, and at an output voltage which is one-tenth of the peak value.

c) Examine the relationship between v_A and v_B near where v_B begins to go positive. Estimate the time (and corresponding phase angle) at which the output voltage is 1/2 the diode drop at the peak, and the corresponding diode drop. Verify your observations by a calculation based on the amplitude and frequency of the input sine wave.

d) Switch the generator to provide a square-wave output. Notice the direct effect of the diode drop.

- **Tabulation:**

    v_A, v_B, v_{Dpeak}, $v_{Dpeak/10}$, $t_{D/2}$, $\Phi_{D/2}$, $v_{D/2}$, v_D.

- **Analysis:**

    Consider the fact that while this circuit demonstrates the overall behaviour of the rectifier quite well, detailed measurements of the diode itself are awkward and inherently imprecise. Various solutions exist for the problem of differential measurement, as noted in Experiment #0 and Appendix B4: The most convenient approach to use is an oscilloscope with a broad-band differential measurement capability to measure the diode drop directly, but this is difficult in practice; Generally speaking, what is needed is more basic knowledge of diodes themselves before we launch into the details of applications. Such is the motive of the Explorations following in E2.1 and E2.2. But first, we will consider one more important system application of diodes where we might gain additional insight into our modelling need.

E1.2 Rectifier Filtering

- **Goal:**

    To explore the use of a capacitor to store energy from a rectifier between intervals of diode conduction, and thereby to **smooth** or **filter** the rectifer output voltage.

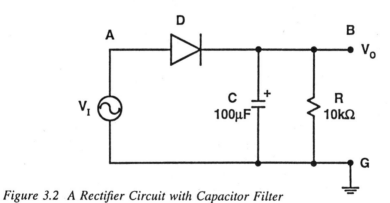

Figure 3.2 A Rectifier Circuit with Capacitor Filter

- **Setup:**

    ○ Assemble the circuit shown in Fig. 3.2. Use a 10 V peak 100 Hz sinewave input. Note that the capacitor $C = 100$ μF is polarized. It *must* be operated with its (+) terminal positive.

- **Measurement:**

    a) Display the waveforms at nodes A and B with a "normalized" oscilloscope. Estimate the diode drop during conduction. Sketch and carefully label both waveforms. Estimate the time interval for which the diode is forward conducting.

b) Shunt R_L by a resistor $R_2 = 1$ kΩ. Measure v_A and v_B as in a).

c) Switch the generator to provide a square wave. Repeat a), b) above with load $R_{eq} = 10$ kΩ or $1k\Omega || 10k\Omega = 0.909$ kΩ.

- **Tabulation:**

    Input type, v_{Apeak}, v_{Bpeak}, $v_{Bvalley}$.

- **Analysis:**

    Consider the advantages of a very large filter capacitor. Calculate the ripple voltage as $v_r = v_{peak} - v_{valley}$ in all cases. What would v_r become if C were doubled? Estimate the fraction of a cycle for which the diode conducts in each case? What is the average output voltage? What is the average diode current?

• DIODE CONDUCTION – THE FORWARD DROP

Now, having seen the diode in action, so to speak, we will explore some basic properties of the most important characteristic of a junction diode, namely its forward drop:

E2.1 Basic Measurements

- **Goal:**

    To explore a simple means of characterizing diode forward drop.

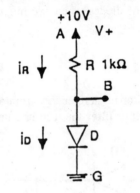

Figure 3.3 A Diode Forward-Drop Test Circuit

- **Setup:**

    ○ Assemble the circuit shown in Fig. 3.3, using a IN4004 diode, a 1 kΩ resistor and a 10 V dc supply. Note that the circuit is essentially the same as that in Fig. 3.1, but with a dc signal source, and the location of circuit ground redefined. [The latter idea, that of choice of an appropriate ground (or common) connection, is an important issue in electronics design (See p. 937 of the Text).]

- **Measurement:** {*Use your DVM (with black lead grounded) to measure nodes A and B.*}

    a) Measure v_B. Adjust the supply to 10.0(0) V (for convenience). Find i_D.

    b) Shunt R with a resistor of equal value (1 kΩ). Measure $v_B = v_D$. Find i_D.

    c) With two 1 kΩ resistors connected, shunt D with a second IN4004 diode (assumed to be matched). What does v_D become? What do you conjecture the current in each diode to be?

- **Tabulation:**

    D#, R, v_B, i_D, for one or two diodes (D) and $R = 1$ kΩ or 0.5 kΩ.

- **Analysis:**

    Consider the significance of the results you have found. Use them and Eq. 3.3 (or 3.4, 3.5) of the Text to find I_S and n of your diodes. At what current level do they have a voltage drop of 0.700 V? How would you refer to them (using the description "x" mA diode)?

E2.2 Diode Measurement with an Ohmmeter[2]

An ohmmeter is quite useful for rapid analysis of diode junctions, in that its equivalent circuit is very much like the circuit of Fig. 3.3. Here we will use it in only a first-order way to reinforce some basic ideas about diodes. Later, we can use it to evaluate transistors. *Note* that the "*resistance*" reading has *no direct significance* relative to the diode characteristic, but makes relative magnitude *comparison* possible directly, and *quantification* possible indirectly:

- Insert some of your diodes into sockets on your prototyping board for convenience of probing with your meter leads. The arrangement is noncritical (as long as they are not shorted), but for convenience and reduction of experimental error, orient their cathodes the same way. A common cathode (or anode) connection (but *not* both!) is OK. With a low ohms scale, measure across one diode. Reverse the meter leads and do it again. What do you conclude about rectification? about the ohmmeter polarity (and its internal battery) and the direction of current flow? **Note** that some ohmmeters have ranges for which the internal voltage is intentionally so low that diodes do not conduct (much) during the measurement of associated resistive networks: Use another range!

- Try another diode type. Note the relative resistance of a large (-area) rectifier diode and a small (-area) signal diode.

- Returning to your first (reference) diode, try other ohmmeter ranges on your DMM.

- Use a second DMM to measure the voltage, V_D, across a diode while its "forward resistance" R is being measured by an ohmmeter. What is the significance of V_D/R?

- **Analysis:**

    Consider the quick study of a diode that an ohmmeter makes possible: You can find its polarity, whether it is open or shorted, its relative junction size, and even points on its forward $i-v$ characteristic (See the procedure immediately above).

E2.3 Forward-Conduction Modelling – FindingA Large-Signal Model

- **Goal:**

    To explore a relatively simple but effective way to characterize a diode over a wide current range.

- **Setup:**

    ○ Return to the circuit in Fig. 3.3, but use four wide-ranging values for R, namely 1 kΩ, 10 kΩ, 100 kΩ, and 1 MΩ and a IN4004 diode. Experimentally, ground the cathode of the diode to be tested, and connect all four resistors to its anode with one end of each open, and to which the 10 V supply will be connected in turn. [Note, from a safety point of view, to limit currents which might flow as a result of accidental misconnection, use the power-supply current-limit feature[3] and/or a small series resistor connected to the supply, say 10Ω or 100Ω.]

---

[2] There is more to be said about ohmmeter measurements of junctions. See Appendix E8 of the first edition of this Manual.

[3] See Experiment #0.

- **Measurement:**
    a) For a supply voltage of 10 V, measure V_D as each resistor is connected to the supply in turn. Approximate the corresponding current. Note that if the supply is raised slightly, to 10.7 or so, the values of current become somewhat "friendlier", being easier to estimate (or calculate) and to plot! Sketch a forward-drop characteristic on a linear current scale, and on a log current scale. From the latter, estimate n.
    b) If time permits, repeat for a second diode type (say a IN914). Plot these results on the graphs above.

- **Tabulation:**
    $i_{D1}, i_{D2}, i_{D3}, i_{D4}, v_{D1}, v_{D2}, v_{D3}, v_{D4}$ for two diodes.

- **Analysis:**
    Consider the following questions: a) If 10 mA is the "normal" operating current, what is the junction voltage at 1% of normal (0.1 mA)? 0.1% (10 µA)? For these values used as V_{DO} (See Fig. 3.20 of the Text), find r_D, b) What is an estimate of incremental resistance at about 5 mA, about 0.5 mA, about 50 µA. What would you estimate it to be at about 50 mA?

E2.4 Forward-Conduction Modelling – Finding A Small-Signal Model

- **Goal:**
    To illustrate a convenient way to characterize a small-signal model.

- **Setup:**
    O Return to the basic circuit described in Fig. 3.3 above.

- **Measurement:**
    a) Connect a IN914 diode using $V^+ = 10$ V and $R = 1$ kΩ. Measure V_D.
    b) Shunt the diode by a resistor which reduces the diode current by 5% to 10%. Notice, in general, that an appropriate value is R. In particular, here use 1 kΩ. Measure V_D. Calculate an estimate for r_d at nearly 10 mA.
    c) Repeat the procedure above for $R = 100$ kΩ, and find r_d at nearly 100 µA.
    d) Repeat all of the above for a IN4004 diode.

- **Tabulation:**
    D, i_D, v_D, v_{DR}, r_d for two diodes and four circuits.

- **Analysis:**
    Consider your results in comparison with the values found using Eq.3.53 of the Text using your estimate of n from E2.3 and a) approximate values of I_D (ie, 10 mA and 100 µA), or b) more precise values of I_D including the effect of the diode-voltage drop and of the shunt.

I hope you enjoyed your taste of diodes: No electronics circuit is possible without at least one!!

EXPERIMENT #4
BIPOLAR-TRANSISTOR BASICS

I OBJECTIVES

Our overall objective is to familiarize you with the basic properties of Bipolar Junction Transistors (BJTs) in preparation for more detailed work on modelling and applications which your future holds. Concentration will be primarily on npn devices for several reasons: Primarily, it will reduce distraction and workload. By concentrating our attention (on only npn devices), other principles may be more easily seen, and comparisons made. As well, npn devices are somewhat special, being more common, having higher gain, and able to operate at higher speed.

II COMPONENTS AND INSTRUMENTATION

Our explorations actually require only one transistor, either a 2N2222, npn, or a 2N3904, npn. However, a second transistor of the same type may be useful in allowing you to verify the relative sensitivity of some circuits to device-to-device variability. As well, a second device may be useful in trouble-shooting. Base diagrams of the transistors are provided in Fig. 4.1. As well, you require a DMM with dc voltage and ohms scales, two power supplies, a two-channel oscilloscope and a waveform generator. As usual, you will use resistors of various values, but primarily 10 kΩ. But since some of the resistors, the 10 kΩ ones in particular, will be used for the rapid evaluation of current amplitudes from voltage measurements, your task will be facilitated if, initially, you use your ohmmeter to measure (and record) values of your resistors.

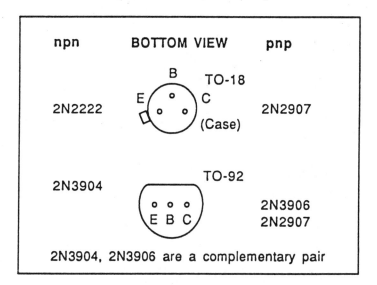

Figure 4.1 BJT Base Diagrams

Obviously, it would be particularly neat and tidy if all 10 kΩ resistors were *exactly* 10 kΩ. However, this is quite difficult to establish in practice. The best you can do is obtain 1% resistors, but even these are generally not identical. Next best is that they all are the same value, whether 10 kΩ or not, but which you can *call* 10 kΩ if you like (with an obvious but small error). Otherwise measure each, and label their values for more precise calculations.

Experiment #4–2

III READING

Reading for this Experiment will be in Sections 4.2, 4.4, 4.7, (4.8), 4.10, and 4.11 of the Text.

IV PREPARATION

As is the norm in these Experiments, **Preparation** is keyed to the **Explorations** to follow by the use of the section numbering employed there.

- ### COMPONENT FAMILIARIZATION AND IDENTIFICATION

    (a) When examined with an ohmmeter, a device, thought to be a BJT, having pins x, y, z provides a set of readings as noted below. Using a labelled diode, the ohmmeter has been found to provide current flowing from its black lead.

Black Lead	Red Lead	Reading
x	y	open
x	z	open
y	x	300Ω
y	z	open
z	x	400Ω
z	y	open

    Identify the transistor type, whether npn or pnp, and which connections are the emitter, collector and base.

P1.1 Establishing Device Currents

(a) In the circuit of Fig. 4.2, using ±10 V supplies, V_B, V_E and V_C are found to be −0.042 V, −0.693 V and +0.697 V respectively. Calculate I_E and the corresponding V_{BE}. Find two estimates each of α and β. In which do you have most faith?

P1.2 Identifying the Controlling Junction and Junction Current

(a) In the circuit of Fig. 4.2, in which the transistor measured above is used, the negative supply is raised to −5 V. Estimate the new values of the 3 electrode voltages under the assumption that $n = 1$.

- ### OTHER, LESS-STABLE BIASING SCHEMES

P2.1 Fixed Base-Emitter-Voltage Biasing

(a) For the circuit shown in Fig. 4.4, with $V^+ = 10$ V, R_p is adjusted until $V_C = 0.697$ V, in which case, $V_B = 0.693$ V. By what amount must the temperature of the transistor rise for saturation at $V_{CEsat} = 0.20$ V to occur? Assume a TC of −2 mV/°C and $n = 1$.

• THE BJT AS AMPLIFIER

P3.1 Voltage Gain and Input Resistance

(a) In the circuit of Fig. 4.5, with R_p adjusted so that $V_C = 5$ V with a 10 V supply, and v_i adjusted to make $v_c = 1$ V peak, estimate the signal v_b. If the signal at node A is three times larger than that at B, estimate R_{inb} and β.

P3.2 Large-Signal Distortion

(a) For the situation described in P3.1 (a), what is the voltage v_c for $v_b = 10$ mV peak? This would normally be the limit of small-signal operation, for which small-signal distortion is a few percent.

(b) For what peak signal values at v_c and v_b does the voltage at node C just reach the edge of saturation.

(c) For what value of peak input signal does the collector current reduce to 1% of its quiescent value.

V EXPLORATIONS

• COMPONENT FAMILIARIZATION AND IDENTIFICATION

It may be instructive to familiarize yourself with your ohmmeter as a means of evaluating properties of junction devices:[1] Use an ohmmeter range whose internal voltage source is high enough (> 0.7 V) to allow a silicon diode junction to conduct. Some ohmmeters intentionally incorporate ranges (often marked with a diode symbol) whose voltages are low enough to allow conventional resistors to be measured even when shunted by diode junctions. Both to verify the ohmmeter polarity and its ability to cause a junction to conduct, check your ohmmeter range(s) using a diode whose banded end is the cathode (from which current flows during conduction). Otherwise, use a second voltmeter to verify the terminal-voltage polarity and magnitude. Finally, realize that larger junctions exhibit (slightly) lower resistances, and that the collector-base junction of a typical transistor is much larger than its emitter-base junction.

- Now make ohmmeter measurements of your transistor(s), thereby checking the transistor type (npn vs pnp) and the pin connections. Note that the "resistance" between collector and emitter is ∞ for either polarity.

Consider the ease with which you can perform this piece of detective work. As noted elsewhere,[2] there is even more for which you can use your ohmmeter in evaluating semi-conductors.

E1.1 Establishing Device Currents (npn)

• **Goal:**
To explore a simple but very effective approach to biasing.

• **Setup:**
○ Note, following the earlier comment, it will make your work much easier if all your "10 kΩ" resistors are the same value (near 10 kΩ), but well-matched (within a fraction of a percent for R_C and R_E, if possible).

○ Connect the circuit as shown in Fig. 4.2, with the supplies *carefully* adjusted to ±10.00 V.

---

[1] More on the issue of ohmmeters is available in Appendix E in Edition 1 of this Manual.
[2] Ibid.

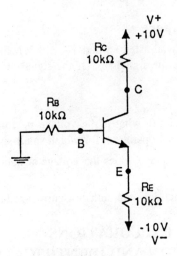

Figure 4.2 A Flexible Biasing Circuit

- **Measurement:**
    a) Measure the voltages at B, E, C with respect to ground, using your DVM. Calculate $V_{BE}, I_E, I_C, I_B, \alpha$, and β, immediately.

- **Tabulation:**
    $V_B, V_C, V_E, V_{BE}, I_B, I_C, I_E, \alpha, \beta$.

- **Analysis**

    Consider the results of your measurements, particularly the values of β and α. Note that unless $R_E = R_C$ within a fraction of a percent, you may find α, calculated as I_C/I_E, to be in error, even appearing to exceed unity! A more appropriate approach to use when resistor values are uncertain is to calculate α as $\beta/(\beta+1)$. Do so, and compare.

E1.2 Identifying the Controlling Junction and Junction Current

- **Goal:**
    To verify the dominant role of the V_{BE} junction in establishing device currents, and to emphasize emitter-current control.

- **Setup:**
    O   Same as in Fig. 4.2, with $V^+ = 10$ V and $V^- = -10$ V.

- **Measurement:**
    a) Raise V^- to –5 V, and measure V_B, V_E, V_C and V_{BE}. Calculate all terminal currents, β and α.

    b) With $V^- = -5$ V, lower V^+ to +5 V, and remeasure and recalculate again.

- **Tabulation:**
    $V_B, V_C, V_E, V_{BE}, I_B, I_C, I_E, \alpha, \beta$.

- **Analysis:**

    Consider what you have learned about the independent and dependent variables involved in transistor current control. Clearly, for active-mode operation (as arranged above), transistor currents depend on conditions in the emitter circuit, and are essentially independent of conditions in the collector. Use equation Eq. 4.3 in the Text and the data above to calculate n^3 and I_S.

- **OTHER, LESS-STABLE BIASING SCHEMES**

E2.1 Base-Current Bias

- **Goal:**

    To demonstrate the inadequacy of a bad, (unfortunately) but common, bias design.

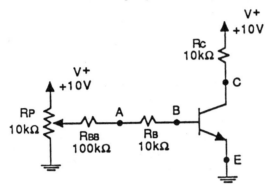

Figure 4.3 A Bad Base-Current-Biasing Circuit

- **Setup:**
    - ○ Connect the circuit as shown in Fig. 4.3.
    - ○ Note that the circuit in Fig. 4.3 is **not** a recommended bias design.

- **Measurement:**

    a) Measure the voltage at node C, adjusting potentiometer R_P until $V_C = +5$ V.

    b) Measure the voltages at nodes A and B with your DVM.

    c) While measuring V_C, heat the transistor, perhaps by blowing through a straw.

    d) Remove the transistor (carefully). Insert another one in its place; Measure V_C.

Tabulation:

V_A, V_B, and values of V_C for two temperatures and two transistors.

- **Analysis:**

    Consider the fact that satisfactory operation of this circuit depends critically on β. As β varies from device to device or with temperature, the voltage V_{CE} will vary greatly, with saturation easily possible for high β. In fact, the very best transistor you can get, one with $\beta = \infty$, does not work at all! This is a clear sign of bad design!

---

[3] See the note on page 230-231 of the Text concerning n.

E2.2 Fixed Base-Emitter-Voltage Biasing

Note, this is clearly the worst bias design of all, unless the desire is to create an electronic thermometer!

- **Goal:**

  To demonstrate the total folly of fixed-voltage bias design.

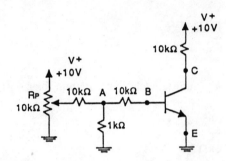

Figure 4.4 A Bad Base-Voltage-Biasing Circuit

- **Setup:**

  O   Connect the circuit as shown in Fig. 4.4.

- **Measurement:**

  a) Adjust potentiometer R_P until the voltage at node C is $V_C = 5$ V.

  b) Measure V_A, V_B with your DVM.

  c) While measuring V_C, heat the transistor by blowing on it through a straw.

- **Tabulation:**

  V_A, V_B, and values of V_C at two temperatures.

- **Analysis:**

  Consider the fact that the base-emitter voltage at a fixed emitter current, drops by 2 mV for each °C rise in temperature. Use your measurement to estimate the rise in temperature you have induced in a burst of expiration!

• THE BJT AS AMPLIFIER

While the circuit shown in Fig. 4.5 uses a rather bad bias design, being a combination of base-current and base-voltage biasing, it is relatively convenient for the measurement of gain of a particular transistor under stable environmental conditions. Incidentally, the presence of the potentiometer R_P is, generally speaking, a sure sign of less-than-ideal design.

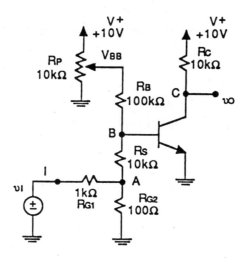

Figure 4.5 A Badly-Biased but Otherwise-Interesting Amplifier

E3.1 Voltage Gain and Input Resistance

- **Goal:**

    To investigate important basic properties of a BJT amplifier.

- **Setup:**

    ○ Connect the circuit as shown in Fig. 4.5. {Note the earlier comment on the suitability of this bias design.}

    ○ Externally trigger your oscilloscope from the waveform generator. Generally speaking, use 10× probes with your oscilloscope, except when you are desperate to view a very very small signal.

- **Measurement:**

    a) With v_i set to zero (or open), adjust R_P so that the dc voltage at C is 5 V.

    b) With v_i a sine wave at 1000 Hz, measure at nodes I and C, adjusting the input-signal amplitude so that v_c is a sine wave of 1 V peak amplitude.

    c) Measure the peak signals at I, A, and B, the latter being quite small, and probably needing a ×1 probe (unfortunately!)

- **Tabulation:**

    v_c, v_b, v_a, v_i, (v_o/v_b, v_o/v_a, v_o/v_i).

- **Analysis:**

    Consider the overall operation of the BJT circuit as an amplifier. Calculate the voltage gains v_o/v_b, v_o/v_a, v_o/v_i, and the current into the base i_b (through R_S) and, thereby, R_{inb}. Note that v_o/v_b is the basic BJT gain while v_o/v_a is the gain which might result from a source whose internal resistance is $R_S = 10$ kΩ. As usual, some signal is lost in the bias network (R_B) although, here, this loss is small, since $R_B >> R_{inb}$.

Experiment #4-8

E3.2 Large-Signal Distortion

- **Goal:**
  To demonstrate that BJT amplifier operation is relatively linear over only a very limited signal range.

- **Setup:**
  - Connect the circuit as in Fig. 4.5.
  - Adjust for $V_C = 5$ V as directed in Exploration E3.1, step a) above.

- **Measurement:**

  a) Measure the voltages at nodes C and I with your dual-channel oscilloscope. Adjust the input voltage so the output is a 1 V peak-to-peak sine wave.

  b) Adjust the oscilloscope channel gain, polarity and dc position controls on the channel connected to node I so that the signals at nodes C and I exactly overlap. You may find it necessary to set the node-C channel on ac coupling,[4] although it is best left on dc coupling, as we shall see.

  c) Raise the input voltage slowly, while observing the voltages at nodes I and C. Note that the output voltage begins to deviate from that at I at the peaks. Note finally (and assuming dc coupling for the channel connected to node C), that the peaks of the output flatten, going no higher than 10 V, nor lower than a few tenths volts above ground. Measure v_b, for the output just noticeably deviating from v_i, and then when it is at its positive peak limit, and then at its negative peak limit.

- **Tabulation:**
  Values of v_C, v_i, v_b at which distortion in v_c is noticed, and then positive and negative peak clipping occur.

- **Analysis:**

  Consider the effects you have observed as evidence of non-linear signal distortion, at first relatively minor, and then seen as very serious clipping as the transistor cuts off and/or saturates (in an order which depends on biasing detail). One normally minimizes distortion by keeping v_b less than 10 mV peak. How do your results compare with this value? {Use your earlier calculation of the gain v_o / v_b.}

We hope that your first hands-on experience with BJTs has been exciting. If so, tell your friends; If not, maintain a dignified silence!

---

[4] See Experiment #0 for more comment on this process of "normalization".

EXPERIMENT #5
MOSFET MEASUREMENT and APPLICATIONS

I OBJECTIVES

The overall objective of this Experiment is to familiarize you, the experimenter, with some of the basic properties of MOS transistors, both n and p channel, and to begin the exploration of some of their fundamental applications.

II COMPONENTS AND INSTRUMENTATION

Our concentration will be on the 4007 MOS array whose package, layout and connections are shown in Fig. 5.1. As can be seen, the array consists of 6 transistors, 3 p-channel and 3 n-channel, interconnected to some extent in order to reduce the number of IC pins required, but otherwise fairly flexible. One critical point to note is that pins 14 and 7, being the substrate connections of all of the p-channel and all of the n-channel devices, respectively, **must** be connected appropriately, no matter what use is made of **any** device. In particular, pin 14 must be connected to the most positive supply in use, and pin 7 to the most negative. Note also that the voltage between pin 14 and pin 7, must be limited to 18 V or so, otherwise internal voltage breakdown may result. **For safety's sake,** maintain this total supply value at or below 16 V.

Other components necessary include resistors of various "unit" values, and a $0.1\mu F$ capacitor, in addition to power-supply filters[1] A second 4007, if available, can be used to evaluate IC-to-IC device variability as an optional exercise.

Instrumentation necessary includes a general-purpose DMM, a dual power supply, a waveform generator and a dual-channel oscilloscope with ×10 probes. A characteristic-curve tracer would be useful, although not essential.

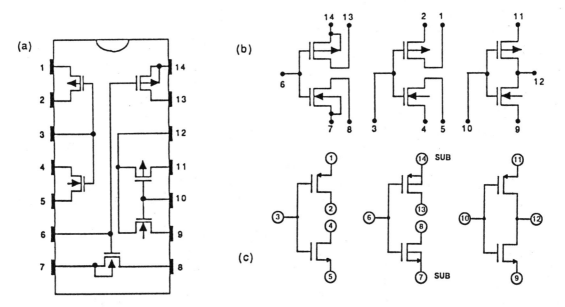

Figure 5.1 The 4007 MOS Array: a) Base Diagram, top view; b) Pin Assignment, substrate-oriented symbols; c) Pin Assignment, simplified

---

[1] See the "Survival Tips" at the end of the Preface, Experiment #0, and Appendix B10.

Experiment #5–2

III READING

The background for this Experiment can be found in Sections 5.1, 5.2, 5.4, 5.5 and 5.6 of the Text.

IV PREPARATION

Following the usual pattern in this Manual, **Preparation** tasks are keyed directly to the **Explorations** to follow, through the use of the same section numbering and titling, but with a P prefix.

- ### DEVICE-INTEGRITY CHECKING

P1.1 Measuring Device Thresholds

(a) For the setup shown in Fig. 5.2, with the supply voltage adjusted to 10.00 V, the DVM reading (node A to ground) is found to be 8.91 V. Estimate the device-threshold magnitude.

P1.2 Measuring the Device Conductivity Parameter

(a) For the setup shown in Fig. 5.2, and the device and situation described in P1.1(a) above, a 1kΩ resistor shunting the DVM reduces its reading to 1.53 V. What value of $k_n = \mu_n C_{ox}(W/L)_n$ applies?

- ### THE AMPLIFIER FUNCTION

(a) For a transistor for which $V_t = 0.90$ V, $k_n = \mu_n C_{ox}(W/L)_n = 0.6$ mA/V^2, in the circuit of Fig. 5.4, calculate the values of V_1 and V_2 required to provide $i_D = 0.1$mA with $v_{DS} = 5$ V.

P2.1 Device Transconductance

(a) For the situation in (a) jsut above, calculate the value of g_m you can expect. For the 10 kΩ drain resistor, what voltage gain would you get? For a sinewave signal at node I of 1 Vpp, what signal would you expect at node C?

- ### THE FEEDBACK-BIAS TOPOLOGY

P3.1 A CMOS Active-Loaded Amplifier

(a) For the circuit of Fig. 5.5, in which $|V_t| = 1.0$ V and $k = 0.6$ mA/V^2, find the device drain currents which result, and V_C.

(b) For $V_A = 30$ V, find g_m, r_o and gain v_d/v_a.

(c) For what peak output and input signals are the output transistors still in saturation?

P3.3 Measuring the Input Resistance

(a) For the situation described in P3.1(a), estimate the input resistance seen to the right of node B. Use the fact (called Miller multiplication) that, for a resistor R connected from output to input of an amplifier of gain A, the corresponding input resistance is $R/(1 - A)$.

(b) For a signal supplied from node I to node A through a 1 MΩ resistor, what is the gain v_d/v_i which results?

V EXPLORATIONS

- ### DEVICE-INTEGRITY CHECKING

The purpose of the two Explorations to follow is to introduce techniques for the rapid evaluation of devices in your MOS array. While you may find the ideas involved not directly obvious at first, depending of course on how much and how thoroughly you have read and understood the Text, the basic underlying process is experimentally very simple, and will become clearer as you proceed. What is most important is that you will practice evaluation techniques which will serve you well later when something does not work as expected, and you wish to verify that your IC still functions.

E1.1 Measuring Device Thresholds
• **Goal:**

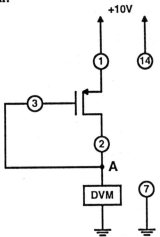

To experience, at first hand, how much useful information can be extracted from a simple experiment (in conjunction with E1.2 to follow); Specifically, here, to find V_t, quickly.

Figure 5.2 Measurement of V_t for a p-Channel Device

• **Setup:**
- O Assemble the circuit as shown in Fig. 5.2 using your array. Ensure that **both** substrate pins (14 and 7) are connected appropriately. All other pins may be left floating. Ensure that the supply is precisely some convenient value (say 10.00 V), by connecting pins 1 and 2 momentarily (and reading the DVM).

• **Measurement:**
- a) Measure the voltage from node A to ground (that is, read the DVM!). Estimate V_{tp}.
- b) Repeat the measurement with drain and source interchanged (ie, pin 2 as source and pin 1 as drain).
- c) Use your DVM to measure V_{tp} of the other p-channel devices having either pins 6 and 13, or 10 and 12, connected as node A, and with pins 11 and 14 joined (with 1 or 2) to the + 10 V supply.

• **Tabulation:**
  Source-pin #, V_A, V_t.

• **Analysis:**

  Consider the ease with which you can measure the important device parameter V_t. Note the degree of device-to-device matching and of device symmetry, you have found.

E1.2 Measuring the Device Conductivity Parameter
• **Goal:**
  To find k, quickly.

• **Setup:**
- O Use the circuit shown in Fig. 5.2, the same as in E1.1, above.

Measurement:

a) Short pins 1 and 2 to measure the supply voltage. Set it to some convenient value (eg, 10.00 V).

b) Now, with the short removed, measure the voltage at the drain-gate common connection (node A).

c) Now, shunt the DVM with resistors of 10kΩ or less until the DVM reading lowers by a reasonable amount, say 1 V. Note that a convenient way to do this is to use a (10kΩ) potentiometer, whose set value is measured subsequently with an ohmmeter. Alternatively, a decade resistor box provides a self-calibrated alternative. Of course, just trying various resistors is OK, since the change does not have to be exact, only measurable. Use the relationship $i_D = 1/2 k_p (v_{GS} - V_{tp})^2$, from Eq. 5.18 in the Text, to find $k_p = k_p'(W/L)$.

- **Tabulation:**

    V_{AO}, R, V_{AR}.

- **Analysis:**

    Consider the ease with which, first V_t, then k are found, using a very simple experimental setup. Note that the relatively complex process used here to measure k, leads to a convenient calculation, but is otherwise *not* essential. As noted above, a measurement with *any* (known) resistor allowing a suitable current flow is all that is needed!

E1.3 Measuring n-Channel Device Parameters

- **Goal:**

    To completely characterize an n-channel device, by finding V_{tn} and k_n, and to do so in a very simple and direct way.

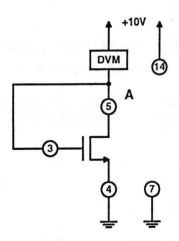

Figure 5.3 Measurement of V_t and k for An n-Channel Device

- **Setup:**

    O   Assemble the circuit shown in Fig. 5.3. Notice the relationship to Fig. 5.2. Note that the NMOS substrate is grounded through pin 7.

- **Measurement:**

    a) First, measure V_{tn}, almost directly, as the difference between the supply voltage and the DVM reading.

b) Then, shunting the DVM by some appropriate resistor, say 1kΩ, find the device voltage, current, and then k_n.

- **Tabulation:**

    R, V (with computed V_t, i_p, and k_n), for $R = \infty$ or 1 kΩ.

- **Analysis:**

    Consider again the ease with which a large amount of information can be found about a MOS device from a very simple experiment.

- **The Amplifier Function** We are now going to explore basic amplifier functions, properties and parameters using the circuit of Fig. 5.4. While this is *not* an appropriate amplifier topology for most applications, it has the advantages in the present context of simplicity with adaptability. The voltage V_1 is a dc supply whose role it is to adjust the dc component of v_{GS} and hence the dc value of i_D. The voltage V_2 is a second dc supply with which one can control the dc value of v_{DS}, once the value of I_D is established. Capacitor C isolates the dc level at node B, but is large enough (compared to the resistance level at B) to represent a short-circuit for signals in the frequency range of interest.

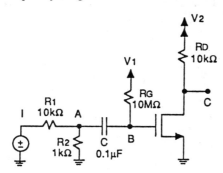

Figure 5.4 A Simple Grounded-Source Amplifier Topology

E2.1 Device Transconductance

- **Goal:**

    To evaluate device transconductance.

- **Setup:**

    ○ Assemble the circuit of Fig. 5.4 using the 345 NFET (the one whose pins are 3,4,5) with substrates connected to ground and V_2 appropriately.

- **Measurement:** {*Use your DVM for steps a) and b) and your oscilloscope otherwise.*}

    a) Adjust the signal generator for zero output. With $V_2 = 6$ V, adjust V_1 until $V_C = 5$ V. Measure V_1 (why not V_B?).

    b) Apply a 1 Vpp sine wave at 1 kHz to node I, while displaying the voltages at nodes A and C on your oscilloscope. Use ac coupling on the channel connected to node C (for these early measurements). Find the voltage gain $A_v = v_c / v_a$ from node A to node C. Under the assumption that $A_v = g_m R_D$, estimate g_m.

- **Tabulation:**

    V_1 I_C, v_b, v_c, v_c / v_b, g_m.

Experiment #5–6

- **Analysis:**

  Consider how this value of g_m corresponds to the calculation using Eq. 5.43 in the Text with estimates of V_t and K you have already made. If you do not have these values, ignore this calculation and carry on.

- **Measurement:**

  c)   With an input signal of zero, raise V_2 to +15 V, and then V_1 until $V_C = 5$ V. Measure V_1 (but not V_B!).

  d)   Now, with a 1 Vpp sine wave at 1 kHz at node I, display and measure the signals on nodes A and C. What is the voltage gain from node A to node C? Estimate g_m.

- **Tabulation:**

  $V_1, I_C, v_b, v_c, v_c/v_b, g_m$.

- **Analysis:**

  Consider how the g_m and bias data you have obtained in this Exploration, can be used to calculate V_t and k of the transistor you are using. {Hint: Think about Eq. 5.45 and 5.43 of the Text.}

E2.2 Signal Distortion

- **Goal:**

  To explore the wave-shape distortion which results with large signals.

- **Setup:**

  ○   Use the circuit of Fig. 5.4 with $V_2 = 15$ V and V_1 adjusted so that $V_C = 5$ V, and a 0.1 Vpp triangle wave at 1 kHz applied to node I.

- **Measurement:**

  a)   Measure nodes A and C, on oscilloscope channel A and B, respectively, with channel B set to ac coupling, and channel A for signal inversion. Adjust the channel A position and gain control channel until the two waveforms exactly overlap.[2]

  b)   Raise the generator output level, until the input and output waveforms differ by 10% in peak amplitude. {If the signals get too large on the screen, you may need to reduce them using the channel attenuators on **both** channels.} Note the peak-to-peak amplitude of the waveform at node C.

  c)   Switch the node-C channel to dc coupling and measure the dc values of the peaks of the triangle waveform.

- **Analysis:**

  Consider the nature of the distortion you are seeing: There are two kinds, one a result of the square-law device characteristic (see Eq. 5.39 of the Text), and the other due to operation in the triode region. What kind do you see? How could you demonstrate the other kind? {Hint: Consider changing V_2.}

- ## THE FEEDBACK-BIAS TOPOLOGY

---

[2] This is a particular kind of channel "normalization" that is very useful in amplifier gain evaluation. See the commentary on "normalization" in Experiment #0.

Our concentration will be on one of the many possible bias techniques that are shown in Fig. 5.39 of the Text, in particular, the one using resistor feedback from drain to gate. There are several reasons for this choice, some practical, some paedogogical: First, this topology allows direct grounding of the source, avoiding the need for a source-bypass capacitor, which must be very large in practice. Second, this topology always guarantees operation of the FET in the saturation region, independent of FET parameters. Third, it extends relatively gracefully from the discrete version to more integrated ones. Fourth, as we shall see, the idea can be generalized to include load biasing as well. Fifth, the resistor from drain to gate directly embodies the idea of feedback, which, as we shall find, is very very important in practical amplifier designs, although often on a more global scale. Finally, though a large resistor is needed, and resistors (particularly large ones) are costly in an IC environment, its value is quite non-critical, making it relatively easier to create with usual IC technologies.

E3.1 A CMOS Active - Loaded Amplifier

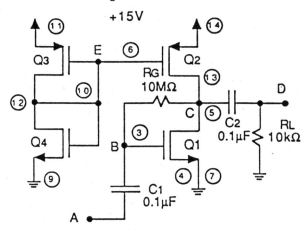

Figure 5.5 A CMOS Common-Source Amplifier

- **Goal:**
  To demonstrate the relatively high gain available from an active-loaded CMOS amplifier.

- **Setup:**
  ○ Assemble the circuit shown in Fig. 5.5 using Q_2, Q_3 and Q_4 to provide a constant-current load to Q_1. Note the substrate connections, *as essential!*

- **Measurement:**
  a) With the supply at 15 V, measure the dc voltages at nodes E and C. Estimate the current in Q_1 and its g_m using earlier measured data.

  b) With a 0.1 Vpp triangle wave at 10 kHz applied to node A, display the waveforms at nodes A, C and D. Calculate the gains v_c / v_a, v_d / v_i.

- **Tabulation:**
  V_c, v_a, v_c, v_d, v_c / v_a, v_d / v_a.

- **Analysis:**
  Consider the relatively large gain you find with an active current-source load. How does it compare with a calculation mode using g_m and R_L? Consider the effect of r_o of Q and Q_2? What is the effect of C_2 on amplifier performance? [This is easier to see if you lower the input

frequency somewhat.]

E3.2 Nonlinear Distortion
- **Goal:**
  To evaluate the effects of signal size on amplifier performance.

- **Setup:**
  - Use Fig. 5.5, as in E3.1 above, with a 0.1 Vpp triangle wave at 10 kHz applied at A.

- **Measurement:**
  a) Superimpose the input and output signals (at A, D) by adjusting the gain of the "node-A" channel appropriately. {Note that you may want to use C as the output node to avoid the frequency dependent effects of C_2, which tend to "bend" the sides of the triangle wave.}

  b) Now, raise the input amplitude, slowly, noting evidence of nonlinearity, then of clipping. At significant events, shift the probe from node D to note the direct-coupled levels at node C.

E3.3 Measuring the Input Resistance
- **Goal:**
  To evaluate the effect on input resistance of the feedback-bias technique.

- **Setup:**
  - Reconsider the circuit of Fig. 5.5, but with a 1 MΩ resistor connected between node A and the generator at node I.

- **Measurement:**
  a) Apply a 0.2 Vpp sine wave at i. Measure the peak-to-peak voltages at nodes I, A, C. Estimate the gain v_c/v_a and the equivalent input resistance R_{in} evaluated to the right of node A.

- **Tabulation:**
  v_i, v_a, v_c, v_c/v_a, R_{in}.

- **Analysis:**
  Consider the value of the input resistance you have found. What value did you expect? What happened? [Hint: Think about the fact that the feedback resistor R_G has v_b at its left and $A_v v_b$ at its right!] Is your confidence shaken or reinforced?

I hope that your first experience with MOS $\underline{T}$ransistors has been MOS$\underline{T}$ enjoyable!

How did you like the idea of active loads? Now you realize that passive loads, like passive people, are rather dull!

EXPERIMENT #6

The BJT DIFFERENTIAL PAIR and APPLICATIONS

I OBJECTIVES

The objective of this BJT experiment is to explore the differential pair, its properties, and potential applications. The general direction taken will be to reinforce the notion of common-emitter half circuits in the process of design and analysis of a differential amplifier.

II COMPONENTS AND INSTRUMENTATION

The major special component needed is the CA3046, a 5-transistor array, whose connection diagram and schematic are shown in Fig. 10.1. It consists of 5 matched npn transistors, two of which share a common emitter connection. In addition, we require a variety of resistors, concentrating on 10 kΩ ± 1% values in order to facilitate the process of matched-pair selection. As well, a 1kΩ potentiometer and two electrolytic capacitors of value 10 µF will be used. Generating and measuring equipment necessary consists of two variable dc supplies (0 to 20 V), a waveform generator, a DMM with 2½ or more digits and ohmmeter ranges, and a two-channel oscilloscope with ×10 probes.

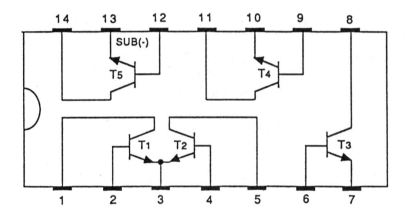

Figure 6.1 The CA3046 Transistor Array

III READING

The material on which this Experiment concentrates, is presented in Sections 6.1 through 6.3 of the Text.

IV PREPARATION

Following the usual pattern in this Manual, **Preparation** tasks are keyed directly to the **Explorations** to follow, by the use of the same titling and section numbering (augmented by the prefix P).

- ## THE BASIC TOPOLOGY
P1.1 DC Parameters and Conditions

(a) For the circuit shown in Fig. 6.2, the following measurements are made: At nodes A and B, − 0.082 V and − 0.079 V; At nodes C and D, −0.771 V and −0.760 V; At nodes E and F, +5.21 V and +5.18 V. The positive and negative supplies are found to be +15.03 V and −10.73 V respectively. Estimate base, emitter and collector currents for both transistors, as well as V_{BE}, β and α, assuming perfect resistors. (Hint: be careful in how you calculate α.)

(b) Estimate the limits on the values of β and α you calculate, for resistor-value deviations of ±1%.

P1.2 Basic Amplification

(a) For the situation described in the first step of E1.2, with $\beta = \infty$, $n = 1$ and $I_E = 1$ mA, what gain would you expect from A to C and from A to E?

(b) For nodes C and D joined, first with 1 kΩ and then directly, estimate the gains, A to C, E, D, F.

P1.3 Input Resistance

(a) For the situation shown in Fig. 6.3, with $\beta = 100$ and $n = 1$, estimate the input resistance to the right of node A with $R_e = 0$ Ω or 100 Ω and with B as shown or grounded. Provide 4 values in all.

P1.4 Loading the Amplifier

(a) For the amplifier of Fig. 6.3, with $\beta = \infty$ and $n = 1$, node B grounded, and $R_e = 100$ Ω, find the gain from A to E, A to F, A to midway between E and F, with no load, and with two 10 kΩ resistors in series as load between nodes E and F.

- **COMMON-MODE REJECTION**

P2.1 Single-Ended Load

(a) For the circuit of Fig. 6.4 with $\beta = \infty$, estimate the common-mode gain.

- **AN IMPROVED CIRCUIT**

P3.1 DC Conditions

(a) For the circuit of Fig. 6.5 with A and B grounded, high β and $V_{BE} = 0.70$ V, estimate the voltages at nodes C, D, E, F, G, H, J, and device currents.

P3.2 Differential Amplification and Common-Mode Rejection

(a) For the situation described in E3.2 following, $v_e = v_f = 0.49$ Vpp, with a 5 Vpp signal on node I, and B grounded. However, v_e and v_f differ by 10 mV with a 5 Vpp signal on node B with I grounded. Estimate the differential and common-mode gains and CMRR (in dB) which apply.

V EXPLORATIONS

- **THE BASIC TOPOLOGY**

Assemble the two circuits shown in Fig. 6.2, using the CA3046 array. Note that in the array, the npn transistors are fabricated on a common substrate whose voltage must be at (or, beyond) the most negative voltage applied to any of the devices. Here it is connected (by pin 13) to the negative supply. The unusual voltage of the negative supply (−10.7 V) is chosen to emphasize the possibility of benefit of design choice, in particular to make the currents nominally 1 mA. Note that it will be convenient if collector emitter, and base resistors are matched in pairs to 1% or better. Furthermore, if their values, as measured by your DVM, are recorded and used in calculations involving α, corresponding results become much more convincing.

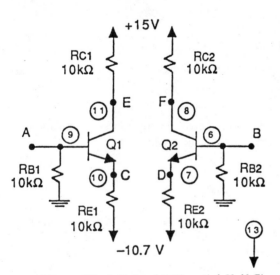

Figure 6.2 A Pair of Differential Half-Circuits

E1.1 DC Parameters and Conditions

- **Goal:**

    To explore the dc aspects of two differential-amplifier half circuits.

- **Setup:**

    ○ Assemble the circuit of Fig. 6.2 using resistors that are as well-matched as you can make them. The better they are matched [between emitter and collector and between half circuits], the more you will learn both about the circuit and the transistors. Use you digital ohmmeter. Make sure your power supplies are bypassed with 0.1 µF ceramic capacitors. Keep you wiring neat with wires short and the two half circuits separated as much as possible.

- **Measurement:** {*While we could obviously measure currents with our DMM (connected as a DCM!), such an approach is awkward, time-consuming, and circuit-disturbing. Furthermore, we do not have enough current meters to make the job even marginally convenient. Correspondingly, we will measure voltages, and use Ohm's Law to find currents.*}

    a) Measure the voltages at nodes A through F and the two power supplies with some attention to precision.

- **Tabulation:**

    V_+, V_-, V_A, V_B, V_C, V_D, V_E, V_F, V_{BE1}, V_{BE2}, $\{I_{B1}, I_{C1}, I_{E1}, I_{B2}, I_{C2}, I_{E2}, \alpha_1, \alpha_2, \beta_1, \beta_2\}$.

- **Analysis:**

    Consider, *now*, the currents in all branches. Do so first, quickly, noting similarities. This process is obviously aided by resistor paired matching. As well, calculate α and β and V_{BE} for each transistor. In general get accustomed to making such estimates "on the fly", so to speak, as part of the "hypothesis and test" aspect of effective dynamic experimentation. (See **Preface** and **Appendix A**.)

- **Measurement:**

    b) Join nodes C and D (which had nearly the same voltage) and measure the voltages on nodes A through F. Note that they are virtually the same as before.

    c) With nodes C and D joined, A open, and B connected via a 1MΩ resistor to the center of a 10kΩ potentiometer, R_p, connected between +15V and −10.7 V, measure the voltage between nodes E and F. Adjust R_p until this is exactly zero. Measure the voltages at A, B, and P, the center of R_p.

- **Tabulation:**

    V_A, V_B, V_C, V_D, V_E, V_F, V_p for two cases.

- **Analysis:**

    Consider what step c) has accomplished: You have in effect compensated for the total input offset including the voltage offset resulting from base-emitter mismatch, and the difference in bias-current flow (ie, offset current) in the base resistors R_B. Estimate the total input offset voltage two ways. What is the average offset current?

E1.2 Basic Amplification

- **Goal:**

    To explore the signal-amplification properties of interconnected differential half-circuits.

- **Setup:**
    ○ For the basic circuit shown in Fig. 6.2, connect node A via a 10kΩ - 100Ω voltage divider to a generator set to provide a sine wave of 1 Vpp at 1 kHz. Ground node B. Note that since both bases are essentially grounded at dc, the dc emitter voltages will be essentially the same (as noted earlier).

- **Measurement:**
    a) Using your two-channel oscilloscope with channel A connected to node I, measure (as carefully as you can), the peak-to-peak voltages at nodes A, C, E. What is the voltage gain from A to C? From A to E?
    b) Now, connect nodes C and D with a 1kΩ resistor, R_e. Measure the peak-to-peak voltages at nodes A through F. What is the voltage gain from A to C, to E, to D, and to F?
    c) Now, repeat the process above but with $R_e = 0$, that is with nodes C and D joined. Measure the peak-to-peak voltages at nodes A, E, F and CD. Find the voltage gains A to E, to CD, and to F.

- **Tabulation:**
    v_a, v_c, v_e, [v_c/v_a, v_e/v_a], v_a, v_b, v_c, v_d, v_e, v_f, [v_c/v_a, v_d/v_a, v_e/v_a, v_f/v_a].

- **Analysis:**

    Consider the fact that the two half circuits, being identical, operate with the same emitter voltages which can be joined with no effect on bias. Rather, only the ac voltage gain is affected by the degree to which each emitter is "grounded" (or has access to ground) through the other emitter. Comment on any interesting relationships between gains that you observe.

E1.3 Input Resistance

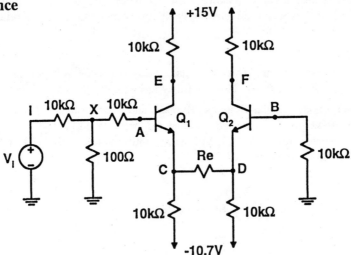

Figure 6.3 A Differential Amplifier Driven by a Single-Ended Signal

- **Goal:**
    To evaluate the input resistance of a differential amplifier.

- **Setup:**
    ○ Modify your circuit (slightly) to the form shown in Fig. 6.3, where a 10kΩ base resistor has been added (at node A). Use a 1 Vpp sine wave at 1kHz for input.

    Note that there are parasitic linkages between devices within the transistor array which can cause the circuit of Fig. 6.3 to oscillate, depending on which transistors in the array are used. For likely success, follow the pin arrangements in Fig. 6.2. {Oscillation is less likely for larger R_e.}

- **Measurement:** {*Since the signal levels at the base are very small, you may have to use ×1 probes to measure them. Generally speaking, while ×1 probes load the measured nodes, and affect oscillation, they are (unfortunately) necessary if your oscilloscope has too little vertical gain.*}

    a) With $R_e = 0\ \Omega$, use your oscilloscope to measure the peak-to-peak voltages at nodes X, A, B, E, F, while maintaining the voltage at node I as a reference.

    b) Repeat with node B grounded.

    c) Repeat with node B grounded, and $R_e = 100\Omega$.

- **Tabulation:**
    B, R_e, v_x, v_a, v_b, v_e, v_f, for two states of node B and two values of R_e.

- **Analysis:**

    Consider what you can learn about the amplifier from the measurements just taken. For example, what is the voltage gain for signal inputs, between nodes A and B, and for outputs measured at E? at F? from E to F? Noting that the voltage difference between X and A allows you to know the input current, estimate the input resistance of the circuit (looking into the base of Q_1) under various conditions?

E1.4 Loading the Amplifier

- **Goal:**
    To explore ways of loading a differential amplifier.

- **Setup:**
    ○ Return to the amplifier of Fig. 6.3, with nodes X and A joined, node B grounded and $R_e = 100\ \Omega$. Use a sine wave input of 1 Vpp at 1kHz. [Note that the 10 µF capacitors, to be used below, are large and bulky, and inappropriate for most IC applications, but provide a convenient way to connect loads to a signal node without disturbing its dc bias. {See, also, Fig. 6.4.}]

- **Measurement:**
    a) With one scope channel connected to node I for reference, measure the peak-to-peak voltages at nodes A, E, F.

    b) Now, load each of nodes E and F with a 10 kΩ resistor coupled by a 10 µF capacitor whose positive end is connected to the collector. Measure the voltages at nodes A, E, F.

    c) Now, directly connect (using no capacitor) the two 10 kΩ loads in series between nodes E and F. Let the connection between them be called node P. Measure the peak-to-peak voltages at nodes A, E, F and P.

- **Tabulation:**
    v_a, v_e, v_f, R_{Le}, R_{Lf}, R_{Lef}, for four arrangements of loading.

- **Analysis:**

    Consider the gains from A to E and F in all cases. Do the results fit with the general idea that gain is proportional to collector resistance? Note that in the last (floating load) case, the load is driven differentially, that is its midpoint is essentially a virtual ac ground. Notice as well that if the load can float, no dc isolation capacitors are needed! For a single 10kΩ load which could be either grounded or isolated, what connection gives the highest gain? What is the value of this gain?

• COMMON-MODE REJECTION
E2.1 Single-Ended Load
- **Goal:**

    To evaluate the response of a differential amplifier to a common-mode signal, one which is applied to both inputs at the same time.

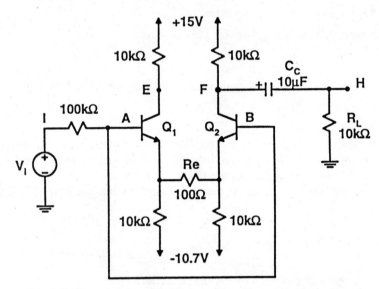

Figure 6.4 A Differential Amplifier Driven by a Common-Mode Signal

- **Setup:**
    - ○ Arrange the circuit with a single-ended load as shown in Fig. 6.4.

- **Measurement:**
    - a) Apply a sinewave signal of 2 Vpp at 1 kHz to the common input. Measure the peak-to-peak signals at nodes I, A, and F.

- **Tabulation:**

    v_i, v_a, v_f.

- **Analysis:**

    Consider both the common-mode gain from A and B to F, and the common-mode input resistance at nodes A and B.

Experiment #6–7

- **AN IMPROVED CIRCUIT**
  - **Goal:**
    Here, the overall goal is to explore the properties of a more IC-like differential amplifier.

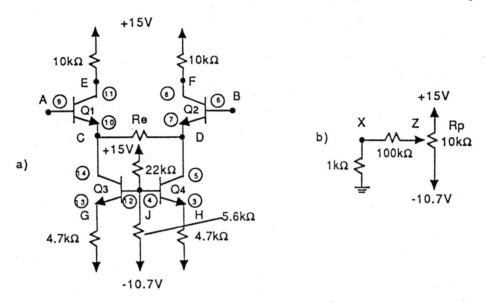

Figure 6.5 An Improved Differential Amplifier

- **Setup:**

  ○ Assemble the circuit as shown in Fig. 6.5. Note that the device substrate (pin 13) is connected to a relatively negative voltage, and is used in a dc part of the circuit. It is best if the two 10kΩ and two 4.7kΩ resistors are pair-matched as well as you can using your ohmmeter. Again, as in E1.3, you may discover that the circuit oscillates, particularly for low values of R_e. However, here, there is little freedom in the choice of transistors. To minimize parasitic coupling between elements, keep the wiring short and neat. If oscillation persists, connect a small capacitor (say 100 pF) from node F to ground. While its effect at 1kHz is relatively minor, it greatly reduces the gain of the parasitic gain paths at high frequencies (1 MHz or more). Again, keep the wiring short, and the two halves of the circuit as separate as short wires will allow. {Note that the leads of a DVM will often cause a high-performance circuit to oscillate, as well. This problem can be avoided by using a resistor (10 kΩ to 100 kΩ) as a "DVM probe".}

Experiment #6-8

E3.1 DC Conditions

- **Goal:**
  To examine the bias conditions and offset voltage.

- **Setup:**
  ○ Assemble the circuit in Fig. 6.5.

- **Measurement:**
  a) With A and B grounded, and $R_e = 1\text{k}\Omega$, use your DVM to measure voltages at nodes C through H and J. Verify circuit operation and general current levels.

  b) Short R_e and measure the voltages at nodes E and F. V_{EF} is the output voltage offset.

  c) Remove node B from ground and connect it to node X of the network shown in Fig. 6.5(b). Adjust R_p until the voltage between nodes E and F is zero. Measure the voltage at node Z. The input offset voltage is about 1% of it.

- **Tabulation:**
  First R_e, V_C, V_D, V_E, V_F, V_G, V_H, V_S, V_{EF}, for two values of R_e; Then V_{EF}, V_B, V_Z.

- **Analysis:**
  Consider the offset voltage situation you have found Compensation of such offsets is an important aspect of circuit design. The scheme introduced above is one you might use with a commercially available op-amp. Alternatively, manufactureers of Op Amp ICs adjust the equivalent of one of the 4.7 kΩ resistors to minimize offset yet leave the inputs (A and B) open and general.

E3.2 Differential Amplification and Common-Mode Rejection

- **Goal:**
  To investigate the gain properties of this differential amplifier including its common-mode rejection.

- **Setup:**
  ○ Augment the circuit in Fig. 6.5 by adding a 1kΩ resistor between nodes A and B, and connecting a 100kΩ resistor to node A, with its free end called I. This will be used for common-mode and difference-mode signal injection.

- **Measurement:**
  a) Now, with $R_e = 1\text{k}\Omega$, ground node B and drive node I with a 5 V pp sine wave signal at 1 kHz. While *externally triggering your scope* from the generator, measure peak-to-peak voltages at nodes A, E, F. Estimate the difference-mode gain for single-ended and differential outputs (and no external load).

  b) Now, reverse the connection: ground I and drive node B with the 5 V pp sine wave at 1 kHz. Again measure nodes A, E, F.

  c) Repeat b) with the signal frequency raised to 100 kHz.

- **Tabulation:**
  f, v_a, v_b, v_e, v_f, for two frequencies and $v_b = 0$ V, or 5 V.

Experiment #6–9

- **Analysis:**

    Consider the fact that the outputs in steps a) and b) above differ only by the effect of common-mode gain. Noting that the difference-mode output signals appear at nodes E and F in opposite phase, while the common-mode signals are in the same phase, estimate the common-mode gain. Calculate the CMRR of this amplifier. Note in step c) the decrease in CMRR with frequency. Can you estimate the location of its 3dB rolloff frequency from the measurements taken (see pages 638-640 of the Text, for interest).

At last, you have seen several ways in which Electronics makes a difference!

NOTES

EXPERIMENT #7
SINGLE-BJT AMPLIFIERS at LOW
and HIGH FREQUENCIES

I OBJECTIVES

The objectives here are two-fold: to explore the particular aspects of the behaviour of capacitor-coupled BJT amplifiers at low-frequencies, and to examine the high-frequency behaviour of the BJT itself, and its operation in simple but classical circuits at high frequencies. The emphasis will be on the latter, high-frequency, part of direct relevance to both discrete-circuit and integrated-circuit (IC) design.

II COMPONENTS AND INSTRUMENTATION

For experimentation, we will concentrate on a discrete BJT, the 2N2222, for which the pin diagram is provided in Fig. 7.1. One 2N2222, only is essential, although a second would facilitate troubleshooting and comparative experimentation, possible if time allows. Alternatively, the npn array CA3046 could be used, for which a pin diagram appears in Fig. 6.1 of Experiment #6. Of course, a 2N3904 would also be a reasonable choice. Necessary components consist of a variety of resistors, dominantly 10 kΩ, of no particular precision, capacitors of 0.1 and 1 µF (low-inductance monolithic ceramic), and 100µF (polarized electrolytic or tantalum), and four mica capacitors of reasonable precision, two 1000 pF and the two 100 pF.

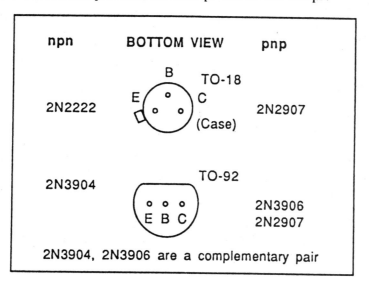

Figure 7.1 BJT Base Diagrams

III READING

Concentration will be on Sections 7.3 through 7.6 of the Text.

IV PREPARATION

As is the recurring pattern in this Manual, **Preparation** will be keyed to the **Explorations** to follow, by the use of the same section numbering and titling employed there, but with a P prefix.

- **THE BASIC COMMON-EMITTER (CE) CIRCUIT**
  (a) Contrast the g_m of a BJT and an MOS device, each operating at 1mA, for $n = 1$ and $|V_t| = 0.5$ V and $k = 10\text{mA}/V^2$.

P1.1 The DC Situation
  (a) For the circuit of Fig. 7.2 with V_S, V_B, V_E, $V_C = -1$ V, -93mV, -770mV and $+5.7$ V, respectively, calculate I_B, I_E, β, α, r_e, r_π, g_m and I_C.

P1.2 Mid-Band Response
  (a) For a sinewave signal of 1 Vpp at 1kHz at node I of Fig. 7.2, estimate the signal voltages at nodes S, B, E, C, D, for the transistor described in P1.1.

- **LOW-FREQUENCY RESPONSE**

P2.1 Basic Overall Response
  (a) For the circuit of Fig. 7.2, with values as in P1.1 above, estimate the critical frequencies and f_{3dB}.

- **HIGH-FREQUENCY OPERATION OF THE COMMON-EMITTER (CE) AMPLIFIER**

P3.1 Basic AC Measurements
  (a) For the circuit of Fig. 7.3, measurements at 10kHz at nodes S, B, C, indicate peak-to-peak voltages of 38mV, 7mV, 2.8V, respectively. Estimate β_{ac}, α, r_e, r_π, g_m, R_i, r_x'.
  (b) The circuit as described in (a) is found to have a 3dB frequency f_1 at 220kHz, with $R_C = 10\text{k}\Omega$, and 420kHz with $R_C = 5\text{k}\Omega$. Estimate C_π, C_μ.

- **(HIGH-) FREQUENCY RESPONSE OF THE COMMON-BASE (CB) AMPLIFIER**

P4.1 Midband (and Low-Frequency) Operation
  (a) For $I_E = 1$mA in the circuit of Fig. 7.4, estimate the gains E to C, and S to C.
  (b) What would you estimate the lower 3dB cutoff, f_L, to be?

P4.2 High-Frequency Response
  (a) For padding capacitors of 100pF and 1000pF applied from the base to the collector and from the base to the emitter, respectively, of the transistor in Fig. 7.4, estimate the upper 3dB frequency f_{H1}. What is the value of the second pole which results?

- **HIGH-FREQUENCY RESPONSE OF THE EMITTER-FOLLOWER (CC) AMPLIFIER**

P5.1 Midband Gain and Upper Cutoff
  (a) For $\beta = 100$ and $I_C = 1$mA, estimate the cutoff frequency, f_{H1}, associated with the input circuit of the "padded" follower.

V EXPLORATIONS

- **THE BASIC COMMON-EMITTER (CE) CIRCUIT**

   Because our emphasis in this Experiment in general will be on high-frequency behaviour, the exploration here of low-frequency behaviour will be relatively limited. In particular, the basic circuit shown in Fig. 7.2, will be emphasized, where only a bypass capacitor and a single output-coupling capacitor are used. The major difference between BJTs and FETs (either JFET or MOSFET) at low frequencies is the overall effect of finite β, and the corresponding impact of components in the base or emitter on what happens in the emitter or base, respectively. We shall experience coupling in one of the directions (only) in what follows. The second important difference between BJTs and FETs (at all frequencies) is the dramatic difference in g_m and therefore in $1/g_m$ (that is of r_e versus r_s). To maintain the large value of gain that high g_m (and low r_e) implies, requires very large emitter-bypass capacitors at low frequencies, as we shall see.

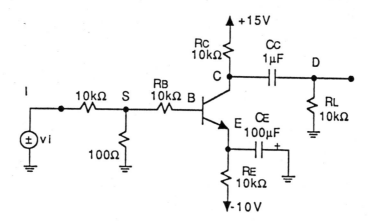

Figure 7.2 A Basic CE BJT Amplifier

E1.1 The DC Situation

- **Goal:**
  To evaluate the dc bias currents and small-signal parameters for the BJT CE amplifier, in preparation for the signal measurement in E1.2.

- **Setup:**
  ○   Assemble the circuit as shown in Fig. 7.2.

- **Measurement:**
  a)   With input grounded, use your DVM to measure dc voltages at nodes S, B, E, C. Use these values to calculate bias currents, β and, thereby, α, r_e, r_π, and g_m. We will assume V_A to be very large, as it normally is, and ignore r_o as a consequence.

- **Tabulation:**
  V_S, V_B, V_E, V_C, I_E, I_C, I_B, β, α.

- **Analysis:**
  Consider the difference between α calculated from β, and directly using I_C and I_E. Why is the latter, the direct path, often so misleading?

E1.2 Mid-Band Response

- **Goal:**
  To evaluate the small-signal response of the CE amplifier at midband, where the coupling capacitors act as short circuits, and the transistor junction capacitors are open circuits.

- **Setup:**
  ○   Use the circuit in Fig. 7.2, with a sine-wave signal of 1 Vpp at 1kHz applied to node I.

Experiment #7-4

- **Measurement:**
  a) Measure peak-to-peak voltages at nodes S, B, E, C, D using your oscilloscope and 10× probes. If your oscilloscope lacks sensitivity, you may use a 1× probe at the low frequency.

- **Tabulation:**
  v_s, v_b, v_e, v_c, v_d.

- **Analysis:**
  Consider the fact that some of the measured voltages are very small: that at E is nominally zero; that at S, while small, is calculable from I; that at B is small, and subject to error, but verifiable using dc data. Use peak-to-peak data to find voltage gains from B to C, and S to D, and the input resistance (looking toward the right at node B).

- **LOW-FREQUENCY RESPONSE E2.1 Basic Overall Response**

  - **Goal:**
    To evaluate the small-signal response of the CE amplifier at midband, where the coupling capacitors act as short circuits and the transistor junction capacitors are open circuits.

  - **Setup:**
    ○ Use the same setup with Fig. 7.2 as in E1.2 above, with a sinewave signal of 1 Vpp at 1 kHz applied to node I.

  - **Measurement:**
    a) With ×10 scope probes on nodes S, D, lower the frequency slowly, identifying first the lower 3dB frequency f_L, then the region in which gain drops by a factor of two per frequency octave, and then by a factor of four per octave, and the intervening break frequency. Finally, try to find the frequency of the transmission zero, below which the response returns to a rate of decline of –20dB/decade. (Incidentally, at what gain from node B to node C does this occur?)

  - **Tabulation:**
    f, v_s, v_d, for $f = f_L, f_B, f_Z$.

  - **Analysis:**
    Consider the situation more thoroughly by sketching a Bode magnitude plot. Verify the critical frequencies you have measured using results from E1.1 and presumed component values. Note that an electrolytic capacitor's value is likely to be quite different than its marked nominal, typically much larger if new and of a good brand, experiment more convenient. A value of 1.0 μF would be more normal.

- **HIGH-FREQUENCY OPERATION OF THE COMMON-EMITTER (CE) AMPLIFIER**

  To make high-frequency measurements on the circuit more convenient and reliable, we will modify the circuit of Fig. 7.2 somewhat, to the form shown in Fig. 7.3:

  In view of the fact that large electrolytic capacitors are inductive at high frequencies, a smaller value of C_E is chosen for which low-inductance monolithic units are available. To avoid spurious effects related to supply wiring inductance, a supply-bypass capacitor C_B is also included. In view of the frequencies to be used, make the wiring relative neat and compact if possible.

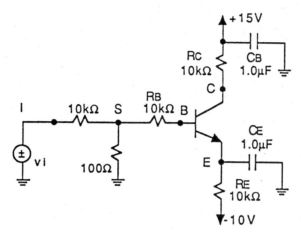

Figure 7.3 Modifications for High-Frequency Operation

E3.1 Basic AC Measurements

• **Goal:**

To characterize the high-frequency response of the modified CE amplifier.

• **Setup:**

○ Assemble the circuit as shown in Fig. 7.3 with a 4Vpp sine wave input at 10kHz. As the frequency is varied in this Exploration, ensure that the input signal amplitude (at I or S) holds constant. Adjust the generator output as necessary.

• **Measurement:**

a) Using ×10 probes and ac coupling, measure the peak-to-peak voltages at nodes S, B and C. Estimate the voltage gains from S to C, B to C and the input resistance looking into the base at B.

b) Now, noting the signal at node C in particular, raise the frequency, until the voltage at C falls by 3dB, to 0.707 of its mid-band value. Note the frequency as f_1.

c) Raise the frequency to $10f_1$, and measure the peak-to-peak voltages at nodes S, B and C.

d) Now, shunt R_C by a second 10kΩ resistor, and proceed to find the modified 3dB frequency, f_2.

• **Tabulation:**

R_C, v_s, v_b, v_c, f, for $f = $ 10 kHz, f_1, $10f_1$, and two values of R_C.

• **Analysis:**

Consider the combination of the results of E3.1 as a means of characterizing your BJT:

a) From the voltages at nodes S, B, C, find β_{ac} and α.

b) From the bias current, find r_e, r_π and g_m.

c) From the ac voltages at S, B, find $R_i = r_\pi + r_x$, and a rough estimate of r_x, ie, $r_x{}'$.

d) From f_1 and f_2 with g_m and R_C, in combination with $(R_B + r_x{}') \parallel r_\pi$, find both C_π and C_μ using the Miller effect for the two gain values.

Experiment #7-6

e) From ac measurements at nodes S, B at $10f_1$, together with values of C_π, C_μ and the gain, find a better value of r_x.

• (HIGH-) FREQUENCY RESPONSE OF THE COMMON-BASE (CB) AMPLIFIER

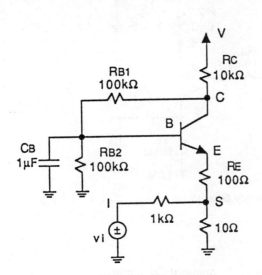

The basic circuit to be explored is shown in Fig. 13.4. The particular topology is chosen to avoid capacitor coupling at the emitter and thereby to simplify measurement. For nominal operation at $I_C = 1\text{mA}$, vary the supply voltage V until the dc voltage across R_C becomes 10 V. R_{B2} is used to ensure adequate signal swing at the collector for high values of β.

Figure 7.4 A CB Amplifier with Feedback Bias and Direct Input Coupling

E4.1 Midband (and Low-Frequency) Operation

• **Goal:**
  To explore midband and low-frequency operation of the CB amplifier.

• **Setup:**
  O   Assemble the circuit shown in Fig. 7.4.
  O   Apply a sinewave signal of 4 Vpp amplitude at 10kHz to node I.

• **Measurement:**
  a) Measure peak-to-peak voltages at nodes S, B, E, C. What is the gain from S to C? E to C?
  b) Lower the frequency of the input signal until the voltage at node C drops by 3dB, at f_L.

• **Tabulation:**
  f, v_s, v_b, v_e, v_c, for $f = 10$ kHz and f_L.

• **Analysis:**
  Consider the significance of the equivalent resistance $R_{eq} = 1/(2\pi f_L C_B)$. What would you have expected f_L to be? Where or what is R_{eq}?

E4.2 High-Frequency Response

As noted in the Text, the common-base amplifier is capable of operation at very high frequencies, typically requiring measurement techniques beyond the budgets and capabilities of the average teaching laboratory. Thus, here, in order even to begin to illustrate high-frequency behaviour, the test circuit will be "padded" with external capacitors intended to model internal behaviour, but at lower frequencies. In particular, in subsequent Explorations, we will shunt the base-emitter junction with a 1000pF capacitor and the base-collector junction with a 100pF capacitor, values whose ratio is representative of that for actual C_π and C_μ, but whose absolute value is about 100 times larger than actual discrete devices, and perhaps 1000 or more times larger than values found for IC ones.

- **Goal:**

    To physically simulate the high-frequency operation of the CB stage at manageable lower frequencies using capacitor "padding".

- **Setup:**
    - ○ Assemble the amplifier in Fig. 74., but with junction capacitances "padded", in particular, with 1000 pF shunting the base-emitter junction, and 100 pF across the collector-base junction.

- **Measurement:**
    - a) With a sinewave input signal of 4 Vpp at 10kHz initially at input I, raise the frequency until the gain measured between nodes S and C drops by 3dB. Note the frequency as f_{H1}.
    - b) Continue to increase the frequency, exploring the "shape" of the frequency response. In particular, explore the limits of the region in which the roll-off is restricted to −20dB/decade.
    - c) With the input at frequency f_{H1} and node C under observation, shunt each padding capacitor in turn by one of equal value, noting any change. Thereby identify the dominating capacitor.
    - d) Now, with the dominating capacitor doubled, find the new 3dB frequency, f_{H2}.

- **Tabulation:**

    $C_{\pi p}$, $C_{\mu p}$, f, v_s, v_c, for a variety of frequencies, first with the standard padding capacitors and then with doubled values.

- **Analysis:**

    Consider what you have discovered about the nature of the common-base cutoff frequency. While Miller multiplication is no longer a problem, the dominating capacitor depends on the stage gain. What capacitor dominates for large values of gain? For what value of gain of the padded amplifier would the two poles coincide? (See Eq. 7.67, 7.68 of the Text.) What new value of R_C would be needed to achieve this gain? Notice of course that it is possible with the low-frequency model using padded capacitors to actually remove one pad at a time and to measure the effect of each alone. Consider this, if you have lots of time.

- ## HIGH-FREQUENCY RESPONSE OF THE EMITTER-FOLLOWER (CC) AMPLIFIER

The circuit we will explore, shown in Fig. 14.5, resembles that used in the original common-emitter (CE) design, except for the addition of padding capacitors (100pF and 1000pF) added for the reasons discussed in E4.0. Since, as noted earlier, electrolytic capacitors can be quite inductive at high frequencies, C_E is shown to consist of a large electrolytic and smaller ceramic capacitor in parallel. Only the ceramic unit is essential for the present purposes (at the frequency range of interest).

Experiment #7-8

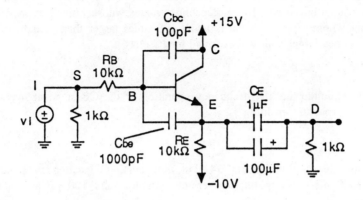

Figure 7.5 A CC Amplifier with Capacitor Padding for Reduced-Frequency Operation

E5.1 Midband Gain and Upper Cutoff

- **Goal:**

    To explore the use of the frequency-scaled padded model in investigating the behaviour of the emitter follower at high frequencies.

- **Setup:**
    - ○ Assemble the circuit as shown in Fig. 14.5, with a 1 Vpp sine-wave input, initially at 10kHz.

- **Measurement:**
    - a) Measure voltages at nodes S, B, E, D, and estimate the gain from S to D and from B to D.
    - b) Raise the input frequency while examining node D, until a 3dB drop in output is observed at f_{H1}.
    - c) Now, with input at f_{H1}, shunt each padding capacitor in turn with one of equal value to identify the dominating capacitor.
    - d) Now, with the doubled dominating capacitor, and input frequency at f_{H2}, shunt R_B by another 10kΩ resistor. Find the new 3dB frequency f_{H2}.

- **Tabulation:**

    $R_B, C_{\pi p}, C_{\mu p}, f, v_s, v_b, v_e, v_d$, for a variety of frequencies, first with normal padding capacitors and then with a doubled dominating value and R_B halved.

- **Analysis:**

    Consider the underlying factors establishing the follower cutoff. Verify some of the results you have found using Eqn. 7.81 of the Text. For this analysis, first employ the padding capacitor values, then the actual values of C_π and C_μ identified earlier in Exploration E3.1. Contrast the cutoff frequency of an emitter follower with that of a common-emitter amplifier.

Experiment #7–9

Since BJTs have the highest transconductance per unit current of any active device, their response at high frequencies is generally acknowledged to be a matter of great importance...., a matter to whose deeper understanding you are privileged to have been guided. You are truly blessed!

NOTES

EXPERIMENT #8
FEEDBACK PRINCIPLES USING an OP-AMP BUILDING BLOCK

I OBJECTIVES

The objective of this experiment is to familiarize you with a sampling of the basic properties of feedback circuits. This process will be conducted at a relatively high level, at which various properties of real circuits (gain, bandwidth, and resistances) can be controlled for purposes of demonstration. The circuit structure employed as a building block will itself be a feedback circuit, but one built on principles available from a very basic knowledge of op-amp techniques. Such as those practiced in Experiment #1 and #2. The concentration will be primarily on the series-shunt feedback topology. But the shunt-shunt topology will be considered as well.

II COMPONENTS AND INSTRUMENTATION

The primary component to be used will be a 741-like op amp, provided in an 8-pin dual-in-line package (DIP) containing two independent amplifiers, and of which two packages, in all, will be used. A block schematic and pin diagram is shown in Fig. 8.1. Two junction diodes, 1N914, are needed as well, along with a collection of resistors and capacitors. In addition, two special capacitors each of 0.1μF and 100μF are needed for power-supply bypassing. Some of the measurements can be made easier and more convincing if some of the 1kΩ, 10kΩ, and 100kΩ resistor values are quite close to nominal (say within 1%).

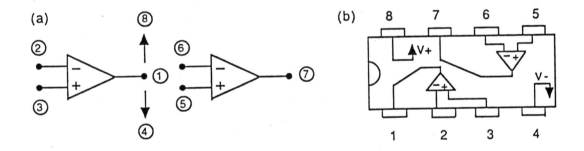

Figure 8.1 Op-Amp Package and Pin Connections

The basic instrumentation required is limited to the usual dual supply, DMM, waveform generator and dual-channel oscilloscope with ×10 probes.

III READING

The scope of this experiment is quite broad, directed (primarily) at a diverse range of functional or systems-level aspects of one of the four feedback topologies, namely the series-shunt circuit. However, a shunt-shunt circuit will be used for the Exploration of multiple-pole loop stability. Accordingly, reading will likewise be broad, covering Sections 8.1 through 8.4, 8.7, 8.10 and 8.11 of the Text.

IV PREPARATION

As is the recurring pattern in this Manual, **Preparation** will be keyed to the **Explorations** to follow, by the use of the same section numbering and titling employed there, but with a P prefix.

- **THE BASIC BUILDING BLOCK**
    (a) Verify the values of μ and f_{3dB} stated in the description of Fig. 8.2. Are they precise? If not, find better analytical expressions.

P1.1 Setup Testing
   (a) For the setup in Fig. 8.3 with $v_i = 1$ Vpp at 100 Hz, what v_c results? What is its 3dB frequency?

- **THE EFFECTS OF FEEDBACK**

P2.1 Finite Gain
   (a) For the circuit as shown in Fig. 8.4, and μ infinite, what is the value of v_c/v_a? What is it for $\mu = 100$ V/V?

R2.2 Finite Bandwidth
   (a) For a 3dB rolloff of μ at 1.6 kHz, what is the cutoff of the closed-loop amplifier in Fig. 8.4 for R_2, changed to 10 kΩ?

R2.3 Output Resistance
   (a) For the closed loop shown in Fig. 8.5, for which $\mu = 100$, estimate the output resistance (looking left at D).

R2.4 Input Resistance
   (a) For the closed loop shown in Fig. 8.6 with $\mu = 100$, find an expression for R_{in} in terms of R. What is it for $R = 100$ Ω?

R2.4 Nonlinear Distortion
   (a) The circuit of Fig. 8.7 has a dc voltage of 0.1 V applied to node A. For $\mu = 100$ V/V and an input offset of 0 V, what voltages appear at nodes B, C, D for a conducting-diode drop of 0.70 volts? Solve this two ways: iteratively or formally. The iterative approach performed mentally is likely to be the most useful one during the heat of laboratory action!

- **MULTIPLE POLES AND LOOP STABILITY**
    (a) Prepare a Bode magnitude plot of A for the situation in which $r = 100$ Ω, $R = 10$ kΩ, $C_1 = 0.1$ μF and $C_2 = C_3 = 0.01$ μF. For what value of β and closed-loop gain is the amplifier stable with a reasonable phase margin? [Hint: See Fig. 8.37 in the Text.]

    (b) What is the phase margin for your choice above? What is the setting of R_p which corresponds?

V EXPLORATIONS

• THE BASIC BUILDING BLOCK

The basic gain block to be employed in many of the Explorations to follow is shown in Fig. 8.2 with the simplifying symbol for it which will be frequently used.

Here A_1 provides the basic voltage gain μ, controlled by resistors R_{1B} and R_{2B}, while A_2 simply buffers the input. As a consequence of A_2, R_{in} is essentially infinite. As a result of the feedback using R_{1B} and R_{2B}, the overall gain $|\mu| \approx R_{2B}/R_{1B}$ (= 100 V/V, for the values shown), and the output resistance is essentially zero. The capacitor C_B shapes the frequency response largely independently of the characteristics of the op amp itself. It provides the basic block with a bandwidth extending from dc to $f_{3dB} = 1/(2\pi R_{2B} C_B)$ (= 1.59kHz for the values shown). Of course, the basic gain block has finite bias currents and an offset voltage. Beyond care in maintaining external resistors below 100kΩ in value, we will tend to ignore such effects. It is obvious however, that FET-Input Op Amps with offset compensation could be used with obvious (but marginal) benefit.

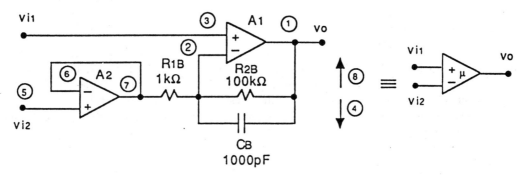

Figure 8.2 A Basic Gain Block

E1.1 Setup Testing

• **Goal:**

To verify the expected behaviour of the constructed basic gain block.

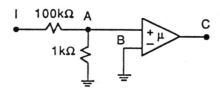

Figure 8.3 A Gain-Block Test Setup

• **Setup:**

○ Assemble the Gain Block as shown in Fig. 8.2, using the pin assignments suggested, with ±15 V supplies. Be sure to include power-supply bypass capacitors (eg, 0.1µF and 100µF) on each supply.

- **Measurement:**

    a) Using the setup shown in Fig. 8.3, apply a 1 Vpp signal at 100 Hz to node I, and measure the peak-to-peak voltages at nodes I, A, C, with your oscilloscope.

    b) Now, with I and C displayed, and compared, load node C with 100 Ω, noting the change in voltage. Your DVM on a sensitive ac range may make this process easier.[1]

    c) Now, with no load, raise the frequency until the voltage at node C drops to 0.707 of its lower-frequency value. Record the frequency as f_{H1}.

- **Tabulation:**

    f, R, v_i, v_a, v_c, for f primarily at 100 Hz with R, initially ∞, then 100 Ω.

- **Analysis:**

    Consider the ideality of your building block. What is its gain? Its output resistance? Its cutoff frequency? Note that its input resistance is difficult to measure, but is very high.

- **THE EFFECTS OF FEEDBACK**

    We are now ready to explore experimentally some of the consequences of feedback circuits on amplifier imperfections. Clearly the basic gain block, our amplifier model in Fig. 8.2, displays finite gain and bandwidth but with each parameter conveniently controllable using single internal components (R_{1B} and C_B respectively). As well, the input and output resistances of the model being nearly ideal (∞ and 0 respectively), they are easily modified (by shunt or series resistors respectively) to demonstrate any desired resistive property.

E2.1 Finite Gain

- **Goal:**

    To explore the effect on closed-loop gain of finite open-loop gain.

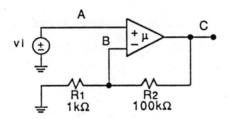

Figure 8.4 A Series-Shunt Feedback Circuits

- **Setup:**

    O   Assemble the series-shunt feedback circuit shown in Fig. 8.4, using the model amplifier μ, with initial values of the β network as shown.

- **Measurement:**

    a) With a 0.1 Vpp signal at 100Hz at node A, measure the voltages at nodes C and B, with both your oscilloscope (using ×10 probes) and DVM.[2] What is the feedback signal, the

---

[1] As noted in Experiment #0, 100 Hz is within the calibrated midband response range of the DVM.
[2] Ibid.

error signal, and the closed-loop gain?

b) Shunt R_2 by a 10kΩ resistor and repeat the process described in the previous step.

c) Short R_2, and repeat the process again.

d) With $R_1 = 1$kΩ and $R_2 = 100$kΩ initially, shunt R_1 by a 1kΩ resistor and repeat the process for the last time.

- **Tabulation:**

  R_1, R_2, v_a, v_b, v_c, for the normal and modified values of R_1, R_2.

- **Analysis:**

  Consider the effect of finite open-loop gain on closed-loop gain, using Eq. 8.4 of the Text. At what value of nominal closed-loop gain (expressed as a fraction of open-loop gain) is the closed-loop gain error (expressed is as a percentage deviation from nominal) 50%? 10%? 1%?

E2.2 Finite Bandwidth

- **Goal:**

  To explore the effect of feedback on closed-loop bandwidth.

- **Setup:**

  ○ Use the circuit as shown in Fig. 8.4, but with $R_2 = 10$kΩ.

- **Measurement:**

  a) With a 1 Vpp sinewave signal at 100Hz applied at node A, measure nodes A and C while raising the frequency slowly. Find the frequency, f_{H2}, where the gain reduces by 3dB.

  b) Repeat the process described above, but with $R_2 = 0$ Ω, to find f_{H3}.

- **Tabulation:**

  R_2, f, v_a, v_c.

- **Analysis:**

  Consider the effect of feedback on closed-loop bandwidth. For this purpose, prepare a Bode magnitude plot of open-loop gain A on which appropriate values of $1/\beta$, and the data concerning f_{H1}, f_{H2}, and f_{H3}, are plotted (where f_{H1} is the 3dB frequency of the Basic Gain Block). Compare with Eq. 8.8 and Fig. 8.37 of the Text.

Experiment #8-6

E2.3 Output Resistance

- **Goal:**
  To explore the effect of feedback on closed-loop output resistance.

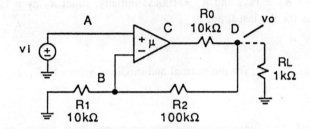

Figure 8.5 A Circuit Which Emphasizes the Effects of Open-Loop Output Resistance

- **Setup:**
  ○ Assemble the circuit shown in Fig. 8.5, with a 0.1 Vpp signal at 100Hz applied to node A, and R_L initially disconnected.

- **Measurement:**

  a) Measure the voltages at nodes A, B, C, D. Find the gains from A to C, and to D.

  b) With nodes A and D displayed, and the gain and offset on the "node-A" channel adjusted for displayed-waveform superposition (normalization[3]), connect the 1kΩ load intermittently. Observe the drop in output peak-to-peak voltage level, and use it to estimate the output resistance of the closed-loop amplifier.

  c) Repeat the process above but with R_2 shorted. [You may find it more convenient to raise the input signal to 1.0 Vpp.] What output resistance now results?

- **Tabulation:**
  $R_1, R_2, v_a, v_b, v_c, v_d$, for two values of R_2, and R_L connected or not.

- **Analysis:**

  Consider the effect of voltage-sensing (shunt-sampling) feedback on the output-resistance-augmented basic amplifier, whose output resistance (seen to the left of node D toward node C), is essentially 10kΩ. Look upon the result as a reduction of a particular disturbance (ie, an increase in output resistance) by a particular factor. What is the factor in general, in terms of μ and the external resistors R_1 and R_2, and for the two specific cases above?

---

[3] See Experiment #0.

E2.4 Input Resistance

- **Goal:**

  To explore the effect of feedback on input resistance.

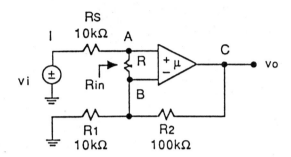

Figure 8.6 Measuring the Effects of Low Input Resistance

- **Setup:**

  ○ Assemble the circuit shown in Fig. 8.6 (using the standard Gain-Block values in Fig. 8.2), with $R = 10\text{k}\Omega$ initially, and a 0.2 Vpp sine wave at 100Hz at node I.

- **Measurement:**

  a) Measure relative voltages at nodes I, A, B (very carefully), and the voltage at node C. [You might find DVM measurements to be useful, and easily possible at 100 Hz.] Estimate the input resistance, R_{in}, of the closed-loop amplifier as "seen" looking right from R_S (into node A).

  b) Repeat the previous measurements with R shunted by a 100Ω resistor.

  c) Repeat the last measurement, with R_2 shorted.

- **Tabulation:**

  R_2, R, DVM readings for v_i, v_a, v_b, for $R = 10\text{ k}\Omega$ and $100\text{ }\Omega$, and $R_2 = 100\text{ k}\Omega$ or $0\text{ }\Omega$.

- **Analysis:**

  Consider the value of R_{in} in relationship to the values of R and the amount of feedback.

E2.5 Nonlinear Distortion

- **Goal:**

  To explore the effects of feedback on nonlinear distortion.

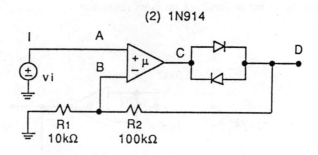

Figure 8.7 Exploring Output Nonlinearities

- **Setup:**
  - ○ Assemble the circuit shown in Fig. 8.7 with an input sinusoid of 0.2 Vpp at 100Hz applied at node I.

- **Measurement:**
  - a) Display the voltages at nodes A and D on your oscilloscope. Adjust the gain and position of the "node-A" channel until the two waveforms overlap.[4] Note the peak amplitudes and other differences, particularly near the zero crossings.
  - b) Display and record waveforms at nodes C and D with the two channels previously "normalized" (to have the same gain, and same average level). Reduce the input-signal level temporarily to 0.02 Vpp, and note the effect.
  - c) Repeat the two previous steps with R_2 shorted, beginning initially with a voltage at I of 0.2 Vpp, comparing A and D, then D and C.

- **Analysis:**

  Consider the effect of the nonlinearity represented by the series diodes. In particular, note the large signal swing at node C which results as the loop attempts to compensate, and the corresponding importance of the value of loop gain. This kind of distortion, called crossover distortion, is discused in Chapter 9 of the Text, and will be explored further in Experiment #9 of this Manual.

- **MULTIPLE POLES AND LOOP STABILITY**

  The remaining Explorations, concerned with various aspects of loop phase shift (and stability) will, like the first ones, operate on the basis of controlled models and simulation, but will use different and more basic components. The specific shunt-shunt feedback circuit to be explored is shown in Fig. 8.8.

  Here, an underlying assumption is that the natural poles associated with each follower are very very high in frequency, and that those associated with the single inverter (without C_1) are high enough to be ignored. With this assumption, the circuit has three controllable poles for which the associated time constants are RC_1, RC_2 and RC_3. The total (open-loop) gain at low frequencies is $-R/r$. Resistors R_1 and R_2 establish a nominal closed-loop minimum gain of -1. Potentiometer R_p allows the closed-loop gain to be adjusted continuously from -1 to $-R/r$.

---

[4] This is the process of "normalization" as discussed in Experiment #0.

Experiment #8–9

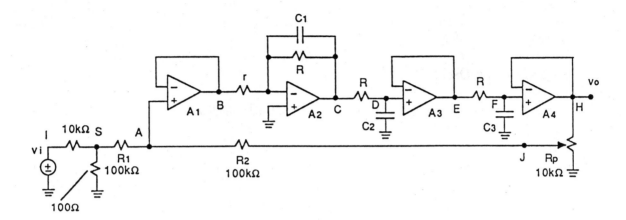

Figure 8.8 A Mutliple-Pole Amplifier having Four Feedback Stages Embedded in a Overall Shunt-Shunt Feedback Loop

E3.1 Stabilization with Three Coincident Poles

- **Goal:**
  To explore the limited feedback possible with a triple pole.

- **Setup:**
  ○ Assemble the circuit shown in Fig. 8.8, with $r = 100\Omega$ and $R = 10k\Omega$, $C_1 = C_2 = C_3 = 0.1\mu F$, and R_p at the bottom of its range. As input at node I, connect a square wave of 1 Vpp amplitude at 1Hz (note the very low frequency) as a way to excite transient behaviour. **Note** that measurements at 1 Hz are usually unnerving! They are a bit more convenient if your oscilloscope is triggered externally, and may suit you more if only a single cycle is displayed.

- **Measurement:**
  a) Display the voltages at nodes S and H.

  b) Adjust the potentiometer R_p carefully (from zero), observing the waveform at node H. For an adjustment at which the overshoot is about 10%, note the physical position of the potentiometer, and measure the amplitude of the flat regions at the top and bottom of the waveforms at nodes S, H, J.

  c) While observing node H, raise the potentiometer tap slowly until oscillation just begins. Note the frequency f_{01}. Measure the peak-to-peak voltages of the oscillation at nodes H and J.

  d) With R_p as adjusted above, shunt R_1 by resistors initially in excess of 1MΩ, and then smaller, to reduce the oscillation at H to a 10% overshoot on the edges of the 1Hz square-wave signal.

  e) Remove power, and using a low-voltage-range ohmmeter, measure the two parts of R_p.

- **Tabulation:**
  $R_P, v_S, v_H, v_J, f, R_{1S}, R_{PH}, R_{PL}$.

- **Analysis:**

    Consider what you have found: Because the amplifier, as set up, has (all) 3 poles coincident and a total phase shift of 270°, it readily becomes unstable for increasing values of β. What is β for sustained oscillation? For 10% overshoot?

E3.2 Loop Gain for Oscillation
- **Goal:**

    To verify untiy loop gain at 0° phase as the oscillation condition.

- **Setup:**

    ○ Use the circuit of Fig. 8.8, as you left it at the end of the previous Exploration. Otherwise, if you are concerned with the R_p setting having changed, repeat E3.1 c) above.

- **Measurement:**

    a) Remove R_1 from node S and grounded it. With R_p at the setting for bare oscillation, remove its upper end (carefully) from the output node (H) and connect it to the waveform generator's input attenuator (node S).

    b) With the generator set to provide a sine wave at node D of 2 Vpp amplitude at frequency f_{01}, present the waveforms of both nodes S and H on your two-channel scope. Check the relative amplitude and phase of the two waveforms as you adjust the generator frequency around the value f_{01}.

- **Analysis:**

    Consider how well you have been able to verify the idea of unity loop gain and phase of 0° at the point of oscillation.

E3.3 Stabilization with Two Coincident Poles
- **Goal:**

    To explore th less restrictive situation with a double pole.

- **Setup:**

    ○ Now, begin again with the basic arrangement in Fig. 8.8 by modifying the circuit to make $C_3 = 10\text{nF}$ while $C_1 = C_2 = 0.1\mu\text{F}$. Adjust R_p to zero, initially, with a 1 Vpp squarewave at 1Hz at node I.

- **Measurement:**

    a) Displaying nodes I and H, increase the setting of R_p. Note the position of R_p for some value of overshoot, say 10%.

    b) At this setting measure the peak-to-peak flat-top parts of the waveform at nodes J and H.

    c) Now, raise the tap on R_p even higher until oscillation just begins. Note the frequency, f_{02}, the peak-to-peak voltages at nodes H and J, and the physical position of R_p.

- **Tabulation:**

    R_p, v_S, v_H, v_J

- **Analysis:**

    Consider the fact that the amplifier with only two dominating coincident poles is easier to stabilize for (lower-gain) closed-loop operation.

E3.4 Stabilization with Staggered and Non-Dominating Multiple Poles

- **Goal:**

    To demonstrate the relative ease of stabilization if the poles are staggered.

- **Setup:**

    ○   Modify the circuit to make $C_2 = 0.01\mu F$, $C_3 = 0.001\mu F$, while $C_1 = 0.1\mu F$, in which case the poles are separated by factors of 10.

- **Measurement:**

    a) Repeat some of the previous, raising R_p to maximize the overshoot. What is the greatest overshoot observed? At what physical setting of R_p?

    b) Now, repeat with $C_1 = 0.1\mu F$, and $C_2 = C_3 = 0.01\mu F$.

- **Tabulation:**

    R_p, v_S, v_H, v_J, for two cases.

- **Analysis:**

    Consider what you have learned about feedback-amplifier stabilization. Note that the the highest values of β and loop gain are possible when the poles are few in number and widely separated!

Feedback can be fun! Believe it or not!

NOTES

EXPERIMENT #9
BASIC OUTPUT-STAGE TOPOLOGIES

I OBJECTIVES

The broad objective of this Experiment is to familiarize you with basic output-stage topologies, of which class-A, class-B and class-AB examples will be explored. In particular, the circuits examined will be BJT versions for various reasons, one being the obvious correspondence with the first part of Chapter 9 of the Text. As well, there is the fact that the low threshold voltages and high g_m of BJTs make them ideal for high-current high-power applications. It is this which motivates the increasing popularity of so-called BiCMOS circuits, which combined these good features of BJTs with the economies of MOS. Another motivating fact is the ready availability of BJT power transistors, of both types, though npn ones are more common in practice.

II COMPONENTS AND INSTRUMENTATION

Here, we will emphasize discrete components, as a possible means for solving special output-stage problems at relatively high power levels. However, here, for convenience, the immediate emphasis will be on relatively-low current and power levels. In particular, we will use two 2N2222 npn BJTs and one 2N2907 pnp BJT, two 1N914 discrete diodes, and resistors of various values – 10kΩ, 1kΩ, and (several) 100Ω. BJT base diagrams are shown in Fig. 9.1. Suitable instrumentation includes two power supplies, a DVM, a waveform generator, and a dual-channel oscilloscope with ×10 probes.

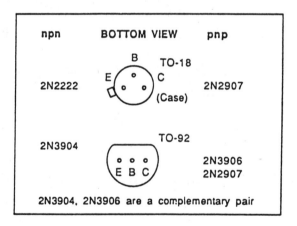

Figure 9.1 BJT Base Diagrams

III READING

Sections 9.1 through 9.5 of the Text are directly relevant to this Experiment. General awareness of BJT operation as presented in Chapter 4 of the Text, will be assumed.

IV PREPARATION

Following the usual pattern in this Manual, **Preparation** tasks are keyed directly to the **Explorations** to follow, by the use of the same titling and numbering. but with P and E prefixes, respectively.

Experiment #9-2

- **THE CLASS-A FOLLOWER**

P1.1 DC Bias

(a) For the circuit shown in Fig. 9.2, with D_1 and Q_2 assumed matched, what current flows in the emitter of Q_1, assuming β to be very high?

P1.2 Signal Operation

(a) For the bias situation in P1.1 (a), but with $\beta_1 = 100$ and $R_L = 10k\Omega$, find the gain v_b / v_s.

(b) What are the upper and lower limits of v_B and corresponding v_A and v_S for a load of $R_L = 1k\Omega$?

- **THE CLASS-B FOLLOWER**

P2.1 DC Bias

(a) For the circuit of Fig. 9.3, what emitter currents flow in each transistor for high β, with $R_L = 1k\Omega$, when $v_S = +3V$, 0V and $-7V$?

P2.2 Signal Operation

(a) For the circuit of Fig. 9.3, sketch the waveform at node B, for v_S a triangle wave of ± 1 volt amplitude, and $R_L = 1k\Omega$. Assume β is high and $|V_{BE}| = 0.7V$.

- **THE DIODE-BIASED CLASS AB OUTPUT STAGE**

P3.1 DC Operation

(a) For the circuit of Fig. 9.4, with $v_I = 0V$, and all junctions matched, what current flows in Q_1 and Q_2? Assume 1mA junctions with a 0.1 V/decade characteristic.

(b) For $\beta = 50$ and $R_L = 1k\Omega$, what is the most-positive output level possible? For what value of v_I does it occur?

P3.2 Signal Operation, at Low and High Current Levels

(a) For very small signals around zero volts, and a $1k\Omega$ load, what is the gain v_h/v_i of the circuit of Fig. 9.4?

(b) For $v_H = V_H + v_h$, with $V_H = 3V$, and $R_L = 1k\Omega$, what value of V_I is required? What is v_h/v_i? Assume $\beta = 50$, $|V_{junction}| = 0.7V$.

- **THE V_{BE}-MULTIPLIER-BIASED CLASS-AB OUTPUT STAGE**

R4.1 DC Operation

(a) For the circuit of Fig. 9.5, with R_p adjusted to the middle, and high β, what current flows in Q_1 and Q_2? Assume $V_{BE} = 0.7V$.

P4.2 High-Power Signal Operation

(a) For the circuit of Fig. 9.5 supplying a 4Vpp square wave signal to a 100Ω load, what input voltage is required? What is the load power, supply power, total device power loss, and efficiency?

V EXPLORATIONS

- **THE CLASS-A FOLLOWER**

In the circuit shown in Fig. 9.2, Q_2 and associated components supply a constant current to the follower Q_1. Resistors R_3 and R_4 serve to equalize the currents in D_1 and Q_2 whose junctions are likely to be quite different in size. A third transistor, Q_3, if available, could be diode-connected to replace D_1. Resistor R_5 serves simply to allow one to monitor the bias current directly. Resistor R_1 represents the internal resistance of a typical signal source.

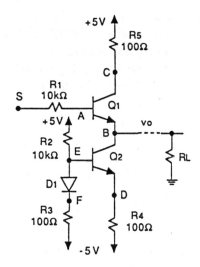

Figure 9.2 A Class-A BJT Follower with Emitter-Current Bias

E1.1 DC Bias

- **Goal:**

    To evaluate a Class-A follower bias design.

- **Setup:**

    ○ Assemble the circuit as shown in Fig. 9.2 with S grounded and no load connected to node B. Adjust the supplies to ±5 V as closely as you can.

- **Measurement:**

    a) Measure the voltages at nodes A, B, C, D, E, F. Estimate the collector current of Q_1 and its β.

    b) Shunt R_2 by a 1kΩ resistor, and remeasure all nodes. What is the bias current in Q_1? What is its β at this current level?

- **Tabulation:**

    R_2, V_A, V_B, V_C, V_D, V_E, V_F, I_C, β, for two values of R_2.

- **Analysis:**

    Consider β and r_e at the current levels you have measured. Notice that the currents in D_1 and Q_2 are nearly the same (due to R_3, R_4).

E1.2 Signal Operation

- **Goal:**

    To evalute the voltage gain and signal-handling capability of the Class-A follower.

- **Setup:**

    ○ Connect the circuit as shown in Fig. 9.2 with R_2 = 10kΩ, a load R_L = 10kΩ, and node S connected to a generator providing a 0.1 Vpp triangle wave at 1kHz.

- **Measurement:**
    a) With your oscilloscope, measure the voltages at nodes S, A, B. Calculate the voltage gains from S to B and A to B, as well as the input resistance at A.
    b) While observing nodes S and B, with both oscilloscope channels direct-coupled with zero volts at the screen center, raise the input voltage until first one output peak, and then the other, limits. At what peak voltages does limiting occur?

- **Tabulation:**
    v_s, v_a, v_b, v_B^+, v_B^-.

- **Analysis:**

    Consider the values of input resistance and gain you find here in comparison with calculations you make using the data of E1.1. As well, correlate the peak limiting values with voltage measurements taken in E1.1.

- **Setup:**
    ○ Reduce the load (R_L) to 1kΩ and apply, initially, a 0.1 Vpp triangle wave at 1kHz.

- **Measurement:**
    c) Measure the peak-to-peak voltages at nodes S, A, B.
    d) While observing nodes S and B, raise the input amplitude until first one peak and then the other peak limit. Note the critical peak voltages at both input and output.

- **Tabulation:**
    R_L, v_s, v_a, v_b, v_B^+, v_B^-.

- **Analysis:**

    Consider the new voltage gain, input resistance, and the nature of the peak output-voltage limitation. Note the relationship between the class-A bias current and the available output swing in one direction!

- **THE CLASS-B FOLLOWER**

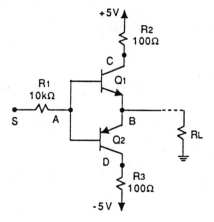

Figure 9.3 A Class-B Complementary BJT Output Stage

Experiment #9–5

E2.1 DC Bias

- **Goal:**
    To evaluate the standing current in the Class-B follower.

- **Setup:**
    ○   Assemble the circuit in Fig. 9.3 with node S grounded, $R_L = 10\text{k}\Omega$, and supplies at $\pm 5\text{V}$.

- **Measurement:**
    a) Use your DVM to measure nodes A, B, C, D.

- **Tabulation:**
    V_A, V_B, V_C, V_D.

- **Analysis:**
    Consider the possibility of standing-current flow.

E2.2 Signal Operation

- **Goal:**
    To observe the dynamic behaviour of the Class-B follower.

- **Setup:**
    ○   As in E2.1 using Fig. 9.3.

- **Measurement:**
    a) With a 0.2 Vpp triangle wave applied to node S initially, and a $10\text{k}\Omega$ load connected to node B, measure nodes S, A and B with your two-channel oscilloscope.

    b) With nodes S and B displayed on a "normalized"[1] oscilloscope, slowly increase the input signal, noting the peak value of the signal at S for which output at node B is just noticeable.

    c) Continue to increase the input amplitude until a 1 Vpp output is observed at B. Sketch and quantify the output. Note the peak input signals corresponding at S and A.

    d) Now, observing node B, increase the input signal until the output peaks just begin to limit. Note their value and the corresponding peak values at nodes S and A. Estimate device β at the peaks.

    e) Now, reduce the input by a few percent from the value used just above and, while displaying node B, examine first node C and then node D on an ac-coupled channel. Note peak values appropriately.

- **Tabulation:**
    $v_s, v_a, v_B, v_S^+, v_A^+, v_B^+, v_S^-, v_A^-, v_B^-$, for various values of v_S.

- **Analysis:**
    Consider what you have discovered: Note the non-ideal performance for small signals (and the open-circuit behaviour for very very small ones). Note the separate roles of the output transistors depending

---

[1] For "normalization", see Experiment #0.

on output polarity and load-current direction. Note that the output voltages are limited essentially by the power supply (and, as we shall see, to a degree by device β and source resistance).

• THE DIODE-BIASED CLASS-AB OUTPUT STAGE

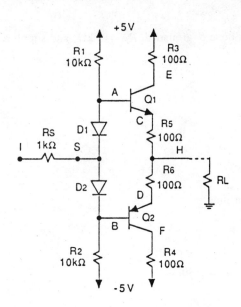

Figure 9.4 A Class-AB Follower

The circuit shown in Fig. 9.4 is a diode-biased class-AB follower, driven symmetrically at the bias-diode common node. This is in contrast to the situation described in the Text (Fig. 9.14) where the class-AB stage is presented as the output stage of a CE amplifier, in which case it is driven from below. Here resistors R_5 and R_6 serve to moderate the current depending on the relative diode and transistor drops. Resistors R_3 and R_4 provide a means by which to conveniently monitor device currents.

E3.1 DC Operation

• **Goal:**

To evaluate the standing current in the Class-AB follower.

• **Setup:**

○ Assemble the circuit in Fig. 9.4 with node I grounded, no load connected, and supplies adjusted to ±5.0 V.

• **Measurement:**

a) With your DVM, measure nodes S, A, B, C, D, E, F, H. Estimate the bias current in Q_1, Q_2.

• **Tabulation:**

$V_S, V_A, V_B, V_C, V_D, V_E, V_F, V_H, I$.

• **Analysis:**

Consider operation of the circuit as a symmetrical mirror. What range of mirror current gains have you observed? Obviously this gain can be controlled by changing the values of R_5 and R_6, or by shunting nodes C to D with a 3rd resistor. What output offset did you find?

E3.2 Signal Operation, at Low and High Current Levels

• **Goal:**

To explore the signal-handling ability of a class-AB follower.

Experiment #9–7

- **Setup:**
    - ○ To the circuit of Fig. 9.4, connect a triangle-wave generator to node I, initially set to 0.2 Vpp at 1kHz, and connect a 10kΩ load to node H.

- **Measurement:**
    - a) Measure the voltages at nodes S, I and H.
    - b) Now, raise the input voltage to find the largest unclipped output available. Measure the peak voltages at I, H.

- **Tabulation:**
    v_s, v_i, v_h, v_I^+, v_H^+, v_I^-, v_H^-.

- **Analysis:**

    Consider the nature of output-signal limitation, in particular the roles of collector voltage, current-equalizing and control components, and device β.

- ## THE V_{BE}-MULTIPLIER-BIASED CLASS-AB OUTPUT STAGE

In the circuit shown in Fig. 9.5, the V_{BE} multiplier, involving Q_3 with R_p, can be used to establish the voltage between nodes A and B, and thereby the current in Q_1 and Q_2.

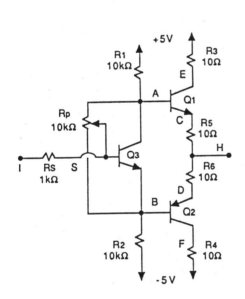

Figure 9.5 The V_{BE} Multiplier in a Class-AB Follower

E4.1 DC Operation *Note: Be very careful* to adjust R_p to near its midpoint *before applying power.*

- **Goal:**

    To verify the standing-current situation for the follower using V_{BE}-multiplier biasing.

- **Setup:**
    - ○ Assemble the circuit shown in Fig. 9.5 with node I grounded, node H open, and the tap on R_p adjusted *to the middle*.
    - ○ Apply power and adjust the supplies as close as possible to ±5 V.

Experiment #9–8

- **Measurement:**
  a) With your DVM, measure node H to verify that it is near zero volts, then measure the voltage across R_5. Adjust the voltage across R_5 to 10mV by varying R_p slightly from center.
  b) Measure dc voltages at nodes S, A, B, C, D, H, E, F, the latter two quite carefully (since their difference from 5.000 V is significant!).

- **Tabulation:**
  $V_S, V_A, V_B, V_C, V_D, V_E, V_F, V_H$.

- **Analysis:**
  Consider the ease with which the quiescent bias current can be adjusted. Of course, in practice, one would normally *not* use a potentiometer, certainly not one with the range of adjustment available here, but rather chose appropriate fixed values, or fixed values with a potentiometer for at most ±10% variability.

E4.2 High-Power Signal Operation

- **Goal:**
  To explore the signal operation of the follower at relatively high current levels.

- **Setup:**
  O Using the circuit of Fig. 9.5, and technique just explored, with I grounded and H open, adjust R_p to produce a 10 mA quiescent current in Q_1 and Q_2, ie, 0.10 volts across R_5 (and also across R_3, R_4, R_6).

- **Measurement:**
  a) Apply a 0.2 Vpp triangle wave at 1kHz to node I and a load of 1kΩ to node H, measuring the peak-to-peak voltages at nodes I, S, H. Find the voltage gain I to H and S to H, as well as the input resistance at node S.
  b) Intermittently connect a 100Ω load (across the 1kΩ) from node H to ground. Measure the peak-to-peak voltage change at node H and estimate the value of R_{out} corresponding.
  c) With a 1kΩ load, raise the input until first one, and then the other of the output peaks begins to limit. Note the corresponding peak values at I, S, H, and calculate the associated gains and input resistance.
  d) Now, with the input at I initially a 0.2 Vpp triangle at 1kHz, and with a 100Ω resistor connected from node H to ground, raise the input signal until the waveform begins to limit, first at one peak, and then at the other. Measure the corresponding peak voltages at nodes I, S, H.
  e) Now, reduce the input to 90% (or so) of the input at the lowest clipping level. Then characterize the interesting aspects of waveforms and peak amplitudes of the signals at nodes $I, S, A, B, C, D, H, E, F$,

- **Tabulation:**
  $R_L, v_i, v_s, v_h, v_I^+, v_S^+, v_H^+, v_I^-, v_S^-, v_H^-$.

Experiment #9–9

- **Analysis:**

    Consider the overall performance of this class-AB output stage in its relatively high-power output mode, by calculating a variety of power output levels, device dissipations and efficiencies. If time permits, contrast these values with ones found below:

- **Measurement:**

    f) Repeat some of the previous steps, but with the standby current adjusted initially to 1mA (rather than 10 mA).

Be sure to note well, and even tell your friends, that while these output stages suffer a [(very) small] voltage loss, they provide a [(very) large] current gain!

Let none of them tell you that a Career in Electronics is Risky. Show them a sure thing: Lose a Little; Win a Lot!

Experiment #9–10

NOTES

EXPERIMENT #10

CMOS OP AMPS

I OBJECTIVES

The objective of this Experiment is twofold: to provide insights into the structure of basic two-stage CMOS amplifiers, and to provide experience with larger-systems applications using a relatively large number of CMOS device arrays.

II COMPONENTS AND INSTRUMENTATION

The primary requirement is for three CD4007 CMOS array ICs. Up to four additional CD4007 can be used for optional enrichment explorations. As well, you need a good collection of capacitors ranging from 10 pF to 0.1µF in a 0.1, 0.33 sequence, a low-inductance low-leakage (ceramic) capacitor of at least 1µF as well as power-supply bypass capacitors. As for equipment, you need a dual power supply, DMM, waveform generator and dual-channel oscilloscope. As well, a capacitor box having 4 or more decades would be useful, though not essential. For convenience, Fig. 10.1 provides various views of the CD4007 package.

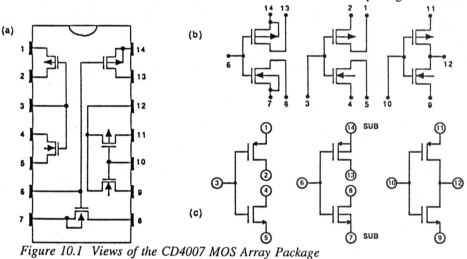

Figure 10.1 Views of the CD4007 MOS Array Package

III READING

Concentration will be on Sections 10.7 and 10.8 of the Text. Familiarity with MOS devices and particularly part of Section 5.7 on CMOS is assumed. Reference will also be made to material on pulse circuits in the latter part of Appendix F of the Text, and to aspects of feedback including resistance modification in Section 8.2, and pole splitting in Section 8.11.

IV PREPARATION

Following the usual pattern in this Manual, **Preparation** tasks are keyed directly to the **Explorations** to follow, by the use of the same titling and section numbering, augmented by the prefix P. Note that there are a large number of preparations here, some of considerable complexity. Clearly, not all can be quickly done. But neither can all of the Explorations! The challenge for you here is to select some of the suggested preparatory investigations to work on. They are very informative, but also time consuming.

- **THE BASIC AMPLIFIER**

P1.1 DC Operation

(a) Consider the CMOS amplifier in Fig. 10.2 employing transistors for which $|V_t| = 1V$, $k = 0.50$ mA/V, and $\lambda = 1/50V$. For ± 7.5 V supplies, $R_1 = 220k\Omega$, $V_F = V_B$, and $V_A = 0V$, estimate all node voltages and device currents, assuming the effect of λ to be negligible.

(b) Now, estimate r_o for each device and the change in bias current that implies.

(c) What difference in the value of V_t for Q_{1B} and Q_{2B} would account for a measured offset of 30mV?

(d) For the bias situation identified in (a) above, find g_m for each device.

(e) Using the value of r_o estimated in (b) and of g_m in (d), estimate the open-loop gain of the amplifier v_f/v_{ab}.

- **AC UNITY-GAIN OPERATION**

P2.1 Over-Compensated Operation with a Dominant Load Pole

(a) Consider the circuit of Fig. 10.2 in which $R_2 = \infty$, $C_2 = 0$ and for which nodes F and B are joined, with a load capacitance, $C_1 = 0.1\mu F$, connected from node F to ground. Estimate its output slew rate for large positive-going and negative-going inputs for which one of Q_{1B} or Q_{2B} is cut off. Also find an approximation to the corresponding 10% to 90% rise and fall times.

(b) Using the values of open-loop output resistance and gain identified in P1.1 (b) and (e), with the relevant value of feedback β, estimate (roughly) the gain and output resistance of the closed loop.

(c) What "high-frequency" 3dB frequency to you expect?

P2.2 Minimal Load-Capacitance Compensation

(a) Consider the circuit of Fig. 10.2 as an open-loop amplifier with a load capacitance $C_1 = 0.1\mu F$. Assume that the amplifier's dynamics are controlled by the output pole and the Miller-Effect-influenced pole at the gate of Q_{6C}, whose C_{gd} is perhaps 2pF, and for which wiring capacitance at node E is possibly 15pF. Estimate the two poles.

(b) Prepare a corresponding Bode magnitude plot of open-loop gain, on which a line for $\beta = 1$ is included. Comment on the stability of the corresponding closed loop.

P2.3 Internal Compensation

(a) Consider the situation in which the output capacitor at node F of Fig. 10.2 is reduced to 100pF, for the conditions in which, for Q_{6C}, $C_{gd} = 2pF$, and C_{stray} (due to wiring) at the gate and drain are perhaps 15pF each. Using Eq. 10.49 and 10.50 of the Text, estimate the location of the corresponding poles.

(b) Prepare a corresponding Bode magnitude plot on which a line for $\beta = 1$ is included. Comment on stability.

(c) For the situation investigated above, to what frequency must the lower pole be moved to ensure stability with 45° phase margin? with 65° phase margin?

- **HIGHER-GAIN OPERATION**

P3.1 An Amplifier with Gain of +100 V/V

(a) For the amplifier of Fig. 10.2 as analyzed in the Preparations above, embedded in the 100 kΩ-10kΩ loop described in E3.1, what closed-loop gain results?

P3.2 Open-Loop Gain

(a) Sketch the circuit described in E3.2, using a triangular amplifier symbol. Calculate the open-loop gain of the amplifier as detailed earlier, which applies at very-low and relatively-high frequencies.

(b) On a Bode magnitude plot of open-loop gain, using the low-frequency value found in (a), with poles as estimated in P2.3 (a) {admittedly with a small load capacitor, a fact we will ignore here}, plot a $1/\beta$ line corresponding to the 10MΩ, 1µF feedback network. Comment on stability. What about the situation with a load resistor of 100kΩ?

• EFFECTS OF DEVICE SIZING
P4.1 Many Possibilities

(a) For one or more of the revised circuits (a), (b), (c) (or any combination) alluded to in E4.1, find the corresponding bias currents and open-loop gain, using the conditions presented in P1.1 (a) above, and ignoring the effects associated with λ.

V EXPLORATION
• THE BASIC AMPLIFIER

The basic amplifier on which you will experiment resembles the two-stage topology shown in Fig. 10.23 of the Text. This is depicted here as well, in Fig. 10.2. Note a major difference here is that choice of device ratios is limited by virtue of array device matching. Initially, the amplifier will be stabilized for low-frequency operation by a large load capacitor.

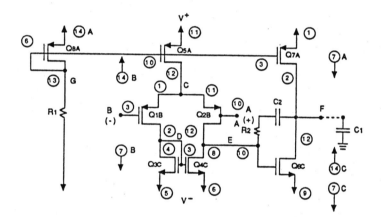

Figure 10.2 A Basic Two-Stage CMOS Op Amp. Three CD4007 arrays (A, B, C) are required. Pin numbers are for the corresponding package. Note the 6 substrate connections, which are essential for correct operation of the arrays

E1.1 DC Operation

• **Goal:**
To verify the DC operation of the assembled CMOS op amp.

• **Setup:**
○ Assemble the circuit shown in Fig. 10.2 using ± 7.5 V supplies, R_1 = 220kΩ, $R_2 = \infty$, and C_2 = 0 pF. Connect the positive input (A) to ground, the negative input (B) to the output (F), and a capacitor C_1 = 0.1µF from output (F) to ground.

• **Measurement:**
a) With zero input voltage, verify that the amplifier is stable by measuring first node F and then node E with your oscilloscope, using a ×10 probe.

b) Using your DVM with a series 10 kΩ resistor as a probe (to minimize various effects of the meter leads), measure (some of the) dc voltages at nodes A through G.

- **Tabulation:**

    V_A, V_B, V_C, V_D, V_E, V_F, V_G.

- **Analysis:**

    Consider the offset voltage you have found. Over what range of input voltages would you expect the output to follow the input reasonably well?

E1.2 Evaluating the Linear Operating Range

- **Goal:**

    To investigate the range of (linear) operation of the amplifier as bias current changes.

- **Setup:**

    ○ Connect the positive input (A) to the center-tap of a 10 kΩ potentiometer (R_3) whose ends are connected to the positive and negative supplies. The negative input (B) remains connected to the output (F).

- **Measurement:**

    a) While measuring the offset voltage directly (between nodes A and B), with your DVM, use R_3 to raise and lower the voltage on node A. For the two settings of pot R_3 at which the offset changes by 0.1 V from its mid-level value, measure the voltages at nodes A, B (and others as you see fit).

    b) Repeat the previous step with R_1 shunted by a resistor, R_4, of equal value. In particular, note the offset voltage for $V_A \approx 0$, and the voltages at nodes A through G at which the offset changes by 0.1 V, and by 0.2 V.

- **Tabulation:**

    R_1, V_{AB}, V_A, V_B, and others, for interesting values of V_A using two offset thresholds.

- **Analysis:**

    Consider the fact that the change in offset voltage at the extremes of input voltage represents the edge of linear operation for large input signals. Try to identify the part of the circuit at which the critical nonlinearity occurs.

• AC UNITY-GAIN OPERATION

E2.1 Over-Compensated Operation with a Dominant Load Pole

- **Goal:**

    To explore the amplifier's operating dynamics.

- **Setup:**

    ○ Assemble the circuit of Fig. 10.2 as indicated in Exploration E1.1 above, but with input A connected to a waveform generator (node I), via a resistor, $R_S = 10$ kΩ.

Experiment #10–5

- **Measurement:**
  a) Using your (normalized) dual-channel oscilloscope, and a 4 Vpp input square wave at 100 Hz, compare the waveforms at A and F. What are the relative amplitudes? What are the times taken for 50% of the total output change? for a change from 10% to 90% of the final output? Sketch the waveforms at A and F.

  b) With input A shunted to ground with a 100Ω resistor, repeat the previous measurements [with an input signal about 1% as large as before].

  c) With conditions otherwise the same as in step b), change the generator input to a sine wave and measure the voltage gain by comparing peak-to-peak values. Now, raise the input frequency until the gain is 0.707 of its lower-frequency value (that is, until it has dropped by 3dB).

- **Tabulation:**
  v_A, v_F, t_{50}, t_{90}, v_a, v_f, f.

- **Analysis:**
  Consider the operation of the circuit as a follower, including estimates of its gain, slew rate, and large- and small-signal bandwidths. Use the relationships in Appendix E of the Text, in particular equation (E.13), to relate rise-times and bandwidths.

E2.2 Minimal Load-Capacitance Compensation

- **Goal:**
  To explore the effects of moving to more minimal compensation.

- **Setup:**
  ○ Use the circuit of Fig. 10.2 as connected in steps b) and c) of Exploration E2.1 above. As well, in preparation for the next step, shunt the load capacitor, $C_1 = 0.1\mu F$, by a second capacitor, $C_{10} = 100pF$, wired with very short leads from output F to ground.

- **Measurement:**
  a) With a 4Vpp square wave at 100Hz applied at the input, note the output waveform in some detail, particularly at the times following slewing where linear operation begins (and continues).

  b) Replace C_1 by a capacitor decade box set to 0.1μF. Again observe the output. For a reasonable box and relatively short connections, you can expect the waveforms to be quite similar to those found previously. If not, you will have to use a selection of discrete capacitors in what follows.

  c) With a 0.1Vpp square wave at 100 Hz at node A, and ×10 probes at nodes A, F, reduce the primary load capacitance C_1 until a peaked oscillatory response is seen at F. Choose a value of C_1 for which the overshoot is some reasonable value (say 10 to 20%). Note the value of C_1 (including the small capacitor C_{10}). Call it C_{12}. (To be somewhat consistent, let $C_{11} = 0.1\mu F$ be the original value of C_1.)

- **Tabulation:**
  Overshoot, C_{12}.

- **Analysis:**
    Consider the possible improvement in dynamics you can expect. By what factor will the slew rate change?

E2.3 Internal Compensation
- **Goal:**
    To explore the possibility of internal-feedback compensation.

- **Setup:**
    o   Use the test setup as established in the last step of E2.2. With C_{12} connected as a combination of discrete capacitors, install R_2 and C_2 as shown in the circuit of Fig. 10.2. Select R_2 initially large (say $R_{20} = 100$ kΩ), with C_2 initially small (say $C_{20} = 10$ pF).

- **Measurement:**
    a) Display nodes A and F using 10× probes. Increase C_2 from C_{20} to C_{21}, a value at which the overshoot at F reduces by 20 to 30% or so from the value established in the last step of E2.2.
    b) With $C_1 = C_{12}$ and $C_2 = C_{21}$, reduce R_2 (by shunting) from R_{20} until the overshoot is minimized, at $R_2 = R_{21}$.
    c) With $C_1 = C_{12}$, $C_2 = C_{21}$ and $R_2 = R_{21}$ initially, change C_1 from C_{12} to C_{13} for which the overshoot is again as large as it was at the end of the last step of E2.2.
    d) Change C_2 from C_{21} to C_{22} in an attempt to reduce the overshoot once more. Then, change R_2 to R_{22} and C_2 again, iteratively, to reduce the overshoot. Call the values finally chosen C_{13}, C_{23} and R_{23}, for convenience.
    e) Now, if time permits, evaluate the effect of changing C_1 from C_{13}, particularly as it is increased.
    f) Now, with the input changed to a 0.1Vpp sine wave, initially at 100 Hz, raise the frequency until the gain falls to 0.707 of its low-frequency value, noting any voltage peaks along the way.

- **Tabulation:**
    C_1, C_2, R_2, overshoot, in many combinations leading ideally to small C_1 and small overshoot.

- **Analysis:**
    Consider the virtues and deficiencies of the unity-gain stabilization process you have just gone through.

- **HIGHER-GAIN OPERATION**

E3.1 An Amplifier with a Nominal Gain of +100
- **Goal:**
    To evaluate operation at increased gain.

- **Setup:**
    o   Connect the circuit of Fig. 10.2 with ± 7.5 V supplies, $R_1 = 220$kΩ, $R_2 = R_{21} = 100$ kΩ, $C_2 = C_{23}$, and load capacitance $C_1 = C_{23}$, or $C_1 = C_2 = 100$pF, if in doubt. Externally, connect a feedback network from node F to node B consisting of a 100kΩ feedback

resistor and 1 kΩ to ground, with a 100kΩ, 1kΩ input-signal divider (with the 1kΩ grounded) connected to the positive amplifier input (A) from the generator (I).

- **Measurement:**
    a) Adjust a square-wave input at 100Hz to provide an output of 1V pp. Measure the peak-to-peak voltage at nodes F and A in order to estimate the closed-loop gain.
    b) Note the output overshoot. Remove the load capacitor (C_1), and note the overshoot again.

- **Tabulation:**
    C_1, v_A, v_F, overshoot.

- **Analysis:**
    Consider your estimate of the closed-loop gain. Note, as we shall verify shortly, that its value is affected by the resistance level of the feedback network. Note also that the loop is more easily stabilized for nominal gains >>1. What evidence do you have for this?

E3.2 Open-loop Gain

- **Goal:**
    To measure the open-loop gain by introducing external dominant-pole compensation.

- **Setup:**
    O Connect the circuit of Fig. 10.2 as described in E3.1 setup, with an input attenuator as indicated, but with a feedback network consisting of a 10MΩ resistor R_f from output F to the negative input B, and a large low-leakage capacitor ($C_3 = 10$ μF, tantalum) from B to ground. Use a sinewave input initially at 10 kHz.

- **Measurement:**
    a) Adjust the input for an output of 1 V pp at F.
    b) Vary the frequency, and note the upper and lower 3dB frequencies.
    c) Assuming a midband region of at least a frequency decade, measure the midband gain. (If the midband is seen to be very narrow, R_f or C_3 must be increased.)
    d) Repeat all of the previous three steps with a resistor $R_L = 1$MΩ connected from the output directly to ground. (If the output offset is large, use a capacitor of 0.1μF in series with R_L.)
    e) Repeat the entire three-step process again with $R_L = 100$kΩ.

- **Tabulation:**
    R_L, v_a, v_f, f_H, f_L with $R_L = \infty$, 1 MΩ and 100 kΩ.

- **Analysis:**
    Consider the open-loop gain you have found, and its dependence on load resistance. Consider also the dependence of the low-frequency cutoff on the value of R_L as it affects the resistance seen by C_3 [See section 8.2 of the Text].

• EFFECTS OF DEVICE SIZING
E4.1 Many Possibilities

While there is limited flexibility available in the control of device sizes when using arrays, one possibility exists. In general, it is to directly parallel array components. The primary precaution to take is to ensure that both the p-channel and n-channel substrates are appropriately connected (to the +ve and −ve supplies respectively).

Several interesting possibilities for parallel connection exist:

(a) Paralleling Q_{1B} and Q_{2B}, with components from a fourth array;

(b) Paralleling Q_{3C} and Q_{4C}, with components from a fifth array;

(c) Paralleling Q_{7A} and Q_{6C}, with components from 2 additional arrays (arrays numbered 4, 5 (or 6, 7 if 4, 5 are already employed in a), b))).

While essentially any of the preceding Explorations can be repeated with such changes, those in E1.1 and E3.2 are quite informative. Obviously, there are a great many possibilities!

CMOS Op-Amps are increasingly important in analog signal-processing systems implemented in Very Large Scale Integration (VLSI).

EXPERIMENT #11
OP-AMP-RC FILTER TOPOLOGIES

I OBJECTIVES

The broad objective of this Experiment is to familiarize you with some of the simpler one- and two-amplifier op-amp filter topologies. The study of two-integrator-loop (three-amplifier) biquads will be deferred to some other occasion, for reasons which include their amplifier count, their systems-oriented (rather than circuit-oriented) nature, the relative directness of their design, and the relative richness and elegance of their application variants. Rather, by concentrating on simpler topologies, emphasis will be in three other directions: toward simplicity, as measured by amplifier count; toward first-order filtering as important, yet straightforward; toward second-order filtering, as expressed through inductance simulation, as a means by which to introduce the rich domain of discrete-component RLC filtering, used at higher frequencies (where reasonably-sized inductors are available). Note that in this experiment there are many many parts; thus *it is very long* and well beyond anyone's ability to complete in a usual laboratory session. It is to provide variety for your enjoyment and education. Please treat it as you would any big bag of tasty treats. Try the ones you fancy. Leave some for another day, to share with others in your Laboratory Class! An interesting subset of Explorations to consider would consist of all or parts of E1 and E4.

II COMPONENTS AND INSTRUMENTATION

As active components, we will use the MC1458P1, a 741-type amplifier, whose pin diagram is shown in Fig. 11.1.

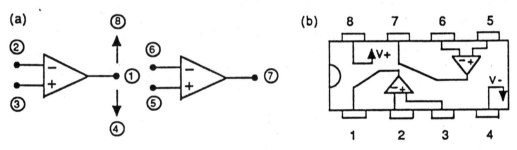

Figure 11.1 Op-Amp Connection Diagram

As well, we will use a small number of passive components, which, as a consequence of the nature of filtering functions in general, should be quite precise, perhaps 1%, or better if available. We will use precise resistors and capacitors as follows: 2 of 1kΩ, 2 of 10kΩ, 5 of 100kΩ, 2 of 0.1μF, 2 of 10nF, 1 of 1nF. If there is a problem in obtaining all of these, the 100 kΩ and 10nF units are the most critical, appearing in situations where pair-matching would suffice, or even be ideal.

Instrumentation required is limited to two power supplies, a waveform generator, and a dual-channel oscilloscope with ×10 probes.

III READING

Background for this Experiment is available in Sections 11.1 through 11.8 of the Text. Emphasis here is primarily on Sections 11.4 through 11.6, and 11.8. Appendix F of the Text provides a useful discussion of STC network response.

IV PREPARATION

As is the recurring pattern in this Manual, **Preparation** will be keyed to the **Explorations** to follow, by the use of the same section titling and numbering employed there, but with a P prefix replacing the E.

- ### A FLEXIBLE OP-AMP TOPOLOGY

P1.1 Mid-Band Gain and Low-Frequency Response

(a) For the circuit shown in P11.2, calculate the midband gain, the very-high-frequency gain, the very-low-frequency gain, and the upper and lower 3dB frequencies.

(b) Prepare a quantified and labelled Bode magnitude plot for the data in (a).

(c) What does your plot tell you about the two zero frequencies?

P1.2 High-Frequency Response

(a) For the circuit of Fig. 11.2, what is the effect on midband gain and upper critical frequencies of reducing R_2 and R_4 by a factor of 10?

P1.3 Impacts of the Need for Higher Closed-Loop Gain on the Filter Frequency Response

(a) What happens to the response of the circuit of Fig. 11.2 when R_2 is increased to 1 MΩ in an attempt to increase the midband gain?

P1.4 Rolloff Compensation – Pole-Zero Cancellation

(a) For C_1, C_2 in Fig. 11.2, left as they are, what is necessary to provide a frequency-independent gain while leaving the complex input impedance unaffected? At what frequency does the amplifier's dominant pole introduce a 3dB rolloff in the compensated closed-loop gain?

- ### SECOND-ORDER TOPOLOGIES

P2.1 The Bridged-T Network

(a) For the circuit of Fig. 11.3, with values shown, find the gain at very high frequencies, and at very low frequencies. What is the frequency of the notch? What is the gain there?

(b) Modify the circuit in Fig. 11.3 by raising R_3 to 1MΩ, and reducing R_4 to 10kΩ. What is the frequency of the notch? What is the gain there? What are the frequencies at which the output is 3dB below the very-high- and very-low-frequency values?

P2.2 Using a Bridged-T Network to Create a Bandpass Filter

(a) Perform the tasks specified in Exercise 11.27 on page 934 of the Text.

- ### SALLEN-AND-KEY CIRCUITS

P3.1 A High-Pass Filter

(a) For the circuit of Fig. 11.5, with values shown, find the transfer function from node F to node C.

P3.2 Other Possibilities

(a) For the circuit of Fig. 11.5, with values shown, find the transfer function from node E to node C.

- **MULTIPLE-AMPLIFIER FILTERS - AN INITIAL GLIMPSE**
P4.1 The Second-Order Bandpass Topology
  (a) For the circuit of Fig. 11.6, with values shown, and driven from node J, calculate the center frequency, the peak gain, the overall Q, and the 3dB and 20dB bandwidths.

  (b) What change is necessary to increase Q by a factor of 2? What change in 3dB and 20dB bandwidths will result?

P4.2 The High-Pass Topology
  (a) For the situation described in E4.2, estimate the high-frequency gain, the peak frequency and amplitude, and both the 3dB and 20dB frequencies.

P4.3 The Low-Pass Topology
  (a) Rework P4.2 for the situation in E4.3.

V EXPLORATIONS
- **A FLEXIBLE OP-AMP TOPOLOGY**

  The circuit shown in Fig. 11.2 is a relatively-simple, yet flexible, first-order topology which allows separate control of both its associated high- and low-frequency single-pole cutoff characteristics. It can be viewed in several different ways, one for example being as a composite of circuits presented in Fig. 11.13 on page 902 of the Text, with additional components. Depending on which circuit(s) in Fig. 11.13 of the Text you consider more fundamental, these components, as well, can be perceived to take on many different roles: For example, R_1 and R_4 can be viewed as making the circuit transmission zeros somewhat more visible. Alternatively, R_1, R_4 can be viewed as arranging to provide non-zero gain at very-low and very-high frequencies. As well, from the point of view of the topology in Fig. 11.13 (c), R_1 can be seen to represent the resistance of a driving source and, accordingly, to make the circuit more realistic.

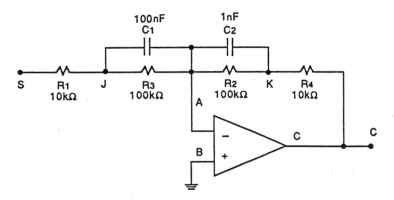

Figure 11.2 A Flexible First-Order Filter Topology

E1.1 Mid-Band Gain and Low-Frequency Response
- **Goal:**

  To explore the low-frequency behaviur of the flexible first-order op amp-RC filter, and to identify its lower pole and zero frequencies.

Experiment #11–4

- **Setup:**
  - ○   Assemble the circuit shown in Fig. 11.2, with ±15V supplies and node S connected to a sinewave generator of 0.2Vpp amplitude at 1kHz.

- **Measurement:**
  - a) Measure the peak-to-peak voltages at nodes S and C, and find the corresponding midband gain.
  - b) Lower the input frequency until the voltage at node C drops by 3dB (ie, to 0.707 of its original (midband) value, at f_1.
  - c) Lower the frequency first to $f_1/2$ and then to $f_1/4$, noting the value of the peak-to-peak voltage at node C, and calculating the gain drop per octave and per decade of frequency change.
  - d) Continue to reduce the input frequency until the gain appears to stop falling, at f_2.
  - e) Setting the generator initially to $f_2/4$, and measuring at nodes S and C, estimate the gain from node S to node C.
  - f) Raise the input frequency slowly until the gain rises by 3dB (to 1.414 of its very-low-frequency value), at f_3.
  - g) Shunt C_1 by a second capacitor of the same value (ie, 100nF). What happens? Now, first lower the frequency until the gain falls to 1.414 times that of its very-low-frequency value, at f_4, then raise it until the gain reaches a value of 0.707 of the midband value, at f_5.

- **Tabulation:**
  f, C_1, v_s, v_c, for various values of f, and two values of C_L.

- **Analysis:**

  Consider the lower frequency-response characteristic you have found. Prepare a suitable Bode plot. [Leave space for the high-frequency characteristic to be found in E1.2.] What is the frequency of the associated pole? Of the zero? What are the very-low and midband gains? What is their ratio? What is the ratio of the pole and zero frequencies (namely f_1/f_3 and f_5/f_4)?

E1.2 High-Frequency Response

- **Goal:**
  To explore the high-frequency behaviour of the first-order op amp-RC filter and to identify its higher pole and zero frequencies.

- **Setup:**
  - ○   With the setup exactly as in Fig. 11.2 (and E1.1 above), change the frequency of the input signal to $4f_1$.

- **Measurement:**
  - a) While measuring nodes S and C, raise the frequency until the gain (and output voltage) drops by 3dB, at f_6.
  - b) Set the input frequency to $2f_6$, $4f_6$ and $20f_6$, in turn, measuring the peak-to-peak output voltages at node C. Estimate the rate of fall of gain for frequencies just above f_6, and well above f_6. As well, calculate the gain at $20f_6$.
  - c) With the input-signal frequency initially at $20f_6$, lower the frequency slowly until the gain increases by 3dB, at f_7.

d) Raise the frequency beyond f_7 again, in an attempt to find a (much much) higher frequency at which the gain again begins to fall, being 3dB down from its value at 10 f_7 (say), at f_8.

e) With the input frequency at f_6, change R_4 to 1kΩ, noting the small change in gain. [By how much?] Then, raise the frequency to detect and verify the new zero frequency, at f_9.

f) With a 0.2 Vpp square wave at 1 kHz (that is, between frequencies f_1 and f_6) applied to node S, find (at C) the output peak-to-peak amplitude, transition times, and sag.

- **Tabulation:**

    f, R_4, v_s, v_C for various values of f and two values of R_4; v_S, v_C, t_t, sag.

- **Analysis:**

    Consider the upper frequency-response characteristic you have found. For the design in Fig. 11.2, what is the frequency of the associated high-frequency pole? Of the zero? What are the midband and very-high gains? What is their ratio? What is the ratio of the zero and pole frequencies?

E1.3 Impacts of the Need for Higher Closed-Loop Gain on the Filter Frequency Response

- **Goal:**

    This optional Exploration considers the effects on the circuit of Fig. 11.2 of an attempt to raise the midband gain by a factor of 10, simply by reducing R_1 from 10kΩ to 1kΩ.

- **Setup:**

    ○ Use the circuit of Fig. 11.2, but with *Rsuv* 1 replaced by a 1 kΩ resistor. To accommodate the increased gain, use a 1kΩ - 100Ω voltage divider feeding node S from the generator (node I).

- **Measurement:**

    a) With a 0.22V sinusoid at frequency $f_7/4$ applied at node I, measure the peak-to-peak voltage at nodes S and C, and note the gain.

    b) Raise the input frequency slowly, noting interesting effects, such as changes in the response rate of change. To better quantify critical frequencies, you may wish to reverse the direction of frequency change in regions of interest. Find the highest measured gain. Over what frequency range is it constant? That is, what is the bandwidth between the corresponding 3dB points?

    c) Explore the behaviour of the circuit above the frequency of the high-frequency zero that you have measured earlier (at f_7). Attempt to identify where the gain begins to fall again at f_{10}.

- **Tabulation:**

    f, v_s, v_c, for notable effects over a wide range of values of f.

• **Analysis:**

Consider what you have observed. Simply put, it is that gain and frequency interact, and that *two components* must be changed to produce only a change in midband gain.

E1.4 Rolloff Compensation – Pole-Zero Cancellation

In the first-order topology of Fig. 11.2, the mutual effects of the two capacitors are decoupled by means of the virtual ground at node A: Specifically, while current flowing in C_2 is controlled by that in C_1, the reverse is not true! Thus, as frequency rises, C_1 in the input network causes both the input and feedback currents to rise, while, independently, C_2 causes the impedance in which the feedback current flows, to fall. In the earlier Explorations, these frequency effects have been intentionally separated to provide a *midband* in which the highest input current flows in the largest feedback impedance to produce the largest output. This exploration considers an interesting special case in which the pole and zero of the input network coincide with the zero and pole of the feedback network, respectively. This occurs for the situation in Fig. 11.2 when $C_1 = C_2$. More generally, complete cancellation is possible when $R_3 C_1 = R_2 C_2$ and $(R_1 \| R_3) C_1 = (R_2 \| R_4) C_2$, or when $R_3 C_1 = R_2 C_2$ and $R_1 C_1 = R_4 C_2$.

• **Goal:**
To explore the occurence of pole-zero cancellation.

• **Setup:**
○   Assemble the circuit of Fig. 11.2, with $C_1 = C_2 = 10\text{nF}$.

• **Measurement:**

a) With a 0.2Vpp sinusoid, initially at 1kHz, applied to node S, measure the peak-to-peak voltages at nodes S and C.

b) Now, explore the response over the range 1Hz to 1MHz.

c) Apply a 0.2Vpp square wave at 1kHz to node S, noting the peak-to-peak amplitude, transition times and sag of the waveform at node C.

• **Tabulation:**
f, v_s, v_c, at several frequencies; v_S, v_C, t_t, sag.

• **Analysis:**

Consider the fact that the response you find (subject to component mismatch and op-amp rolloff) is frequency independent.

• SECOND-ORDER TOPOLOGIES

Here, we will examine the means by which virtually the same components as those already used, can achieve a much-more-complex second-order response. The techniques are interrelated by virtue of the use of the recurring idea of "multiplicity": In passive T networks, (such as that in Fig. 11.3) the "multiplicity" is of signal paths through which signals proceed in parallel, offering the possibility of signal interference or cancellation. In various feedback networks, a diffuse view of "multiplicity" can be considered to exist in the number of times a signal can traverse a loop, during which endless process some attributes are reinforced and emphasized over others.

Experiment #11–7

E2.1 The Bridged-T Network

• **Goal**

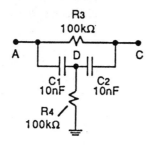

To explore the interesting frequency behaviour of an isolated Bridged-T circuit, and thereby appreciate the effect of signal-path "multiplicity".

Figure 11.3 The Bridged-T Circuit

• **Setup:**

    ○   Assemble the passive Bridged-T network shown in Fig. 11.3, with node A connected to a sinusoidal signal of 0.2Vpp at 10Hz. [Note that your results will be more impressive as components are better selected to be closer to their ideal nominal value.]

• **Measurement:**

    a)   Measure the peak-to-peak output at nodes A and C, and estimate the network gain. Be sure to use ×10 probes!

    b)   Raise the frequency slowly, noting the frequency at which the output drops by 3dB, at f_1.

    c)   Continue to raise the frequency to find that frequency at which the output is a minimum, at f_2. Measure the gain there.

    d)   Above f_2, find the frequency at which the gain is 0.707 once again, at f_3.

• **Tabulation:**

    f, v_a, v_c, for significant values of f.

• **Analysis:**

Consider this interesting behaviour of a passive circuit. As noted earlier, it results from an interference between signals in a multiplicity of paths, that is, anti-phase-signal cancellation at a particular frequency. What is the peak loss in dB? What is the 3dB bandwidth?

• **Setup:**

    ○   Modify the circuit shown in Fig. 11.3, to have R_3 = 200kΩ (using two 100kΩ resistors in series), and R_4 = 50 kΩ (using two 100 kΩ in parallel).

• **Measurement:**

    e)   Repeat the transfer-characteristic measurements above to find the 3dB frequencies at f_4, f_6, and the null frequency, at f_5.

    f)   Shunt one of the series 100 kΩ resistors by 1 MΩ and reevaluate the new minimum, at f_7, and the gain there.

    g)   With additional components removed, change the waveform of the input signal to a square wave, adjusting its frequency, in turn, around the following values: f_5, $f_5/2$, $f_5/3$, $f_5/4$, $f_5/7$, for the most interesting effects.

h) If time permits, it may be very interesting to repeat some of the exploration steps above, with $C_1 = C_2 = $ 10nF as before, but with R_3, R_4 raised and lowered, respectively, by factors of 10, to $R_3 = $ 1MΩ and $R_4 = $ 10kΩ. Use components which are as precise as you can arrange, in order to see the best effect!

- **Tabulation:**

    f, v_a, v_c, for significant values of f.

- **Analysis:**

    Consider the ease with the bandwidth of the notch can be controlled, although the notch depth is a sensitive function of component matching. As noted in g) above, you can use this notch circuit to eliminate a particular signal component with only a modest effect on other signals in the pass band. One application of a high-Q T is in the elemination of spurious 50/60Hz (power frequency) interference in measurement systems.

E2.2 Using a Bridged-T Network to Create a Bandpass Filter

When a Bridged-T network is used to provide feedback in an op-amp circuit, the closed-loop gain is largest at the network's transmission null. Correspondingly, the peak gain, being inversely related to the depth of the null, can be highly component-tolerance-dependent. The basic feedback circuit we will examine is shown in Fig.11.4, where the multiple 100kΩ resistors allow us to use available precise units, as well as providing some experimental flexibility. We will apply inputs to various normally-grounded components through a 10kΩ - 100Ω voltage divider, with input node I and output node S.

- **Goal:**

    To explore the flexibility of a Bridged-T feedback network, in creating bandpass filters.

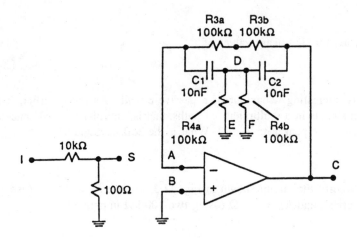

Figure 11.4 An Amplifier with Bridged-T Feedback

- **Setup:**

    O Assemble the circuit in Fig. 11.4. Remove node B from ground, and connect it to node S, while I is driven by a sinusoidal waveform of 1Vpp at 1kHz.

- **Measurement:**

    a) Measuring at nodes $S(=B)$ and C, lower the frequency to find the frequency f_1 at which maximum output occurs. Adjust the input-signal amplitude as necessary. Estimate the peak gain.

    b) First lower, then raise, the frequency to identify the frequencies at which the gain drops by 3dB (at f_2, f_3).

- **Tabulation:**

    f, v_s, v_c for various frequencies.

- **Analysis:**

    Consider the circuit as a bandpass amplifier. What is its midband gain, the 3dB bandwidth, the high- and low-frequency losses? Contrast the behaviour found with that of the basic network itself (in E2.1 above).

- **Setup:**

    O Now, ground node B, unground E, and connect it to S (of the input divider). Set the input I to a sinewave of 2Vpp amplitude at frequency f_1 (identified in the step above).

- **Measurement:**

    c) Measuring the peak-to-peak voltages at E and C, adjust the frequency slightly to ensure maximum gain. Note the gain, and the frequency f_{11}.

    d) Vary the frequency to identify the (lower and upper) 3dB and 20dB frequencies (f_{12}, f_{13}, f_{14}, f_{15} respectively).

    e) Lift R_{4b} from ground, as well, and connect it (ie, both nodes E, F) to node S, and repeat some elements of the last two steps (identifying f_{21}, f_{22}, f_{23}, f_{24}, f_{25}).

    f) Disconnect R_{4b} completely (leaving node F open), and short R_{3b}, repeating all the earlier processes to identify midband gain, center frequency, and bandwidths (with f_{31}, f_{32}, f_{33}, f_{34}, f_{35}).

    g) With B grounded and input at E, switch the input signal to a square wave of 2Vpp amplitude at frequency $10f_{11}$, initially. Now, while observing node C, lower the frequency, noting the output for input signals at $10f_{11}$, $4f_{11}$, $2f_{11}$, f_{11}, $f_{11}/2$, $f_{11}/3$, $f_{11}/4$, $f_{11}/5$, $f_{11}/7$.

- **Tabulation:**

    f, v_s, v_c, for several cases and many frequencies.

- **Analysis:**

    Consider this bandpass filter, specifically its gain and bandwidth, relative to the first one in E2.2, and with respect to the theory presented on pages 930 to 934 of the Text.

• SALLEN-AND-KEY CIRCUITS

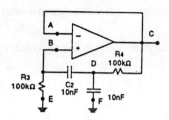

These popular second-order circuits, referred to as single-amplifier biquads (or SABs for short), can be derived from various T-network feedback topologies, as indicated by example on pages 935 to 938 of the Text. Note the interesting fact that positive feedback is used with a unity-gain buffer. The specific variant we will explore is shown here in Fig. 11.5, with input nodes left unspecified.

Figure 11.5 A Single-Amplifier Biquad Building Block

E3.1 A High-Pass Filter

• **Goal:**

To explore the behaviour of a Sallen and Key SAB high-pass circuit.

• **Setup:**

○ Assemble the circuit as shown in Fig. 11.5, with node F removed from ground and node I connected to a 2Vpp sinusoid at 1kHz.

• **Measurement:**

a) While measuring nodes I and C, lower the frequency, noting the frequency f_1, at which the output falls to 0.707 of its high-frequency value, and then the gain at $f_1/2$ and $f_1/10$.

• **Tabulation:**

f, v_i, v_c, for interesting frequencies.

• **Analysis:**

Consider the low-frequency rolloff in view of Eq. 11.73 through 11.76 on page 931 of the Text. What is the 3dB frequency, and the rolloff in dB/octave and dB/decade?

• **Setup:**

○ Now, add a another 100kΩ resistor in series with R_3, and yet another in parallel with R_4.

• **Measurement:**

b) Repeat the previous step, noting, as well, the value and location of any response peak.

c) To the modified filter apply a square wave of 2Vpp amplitude at 1kHz, measuring node C. Sketch the output waveform, noting sag and overshoot. Repeat for inputs at 100Hz and 10kHz.

• **Tabulation:**

f, v_i, v_c, for interesting frequencies.

• **Analysis:**

Consider the overall effect of change of Q on various aspects of the response.

Experiment #11-11

E3.2 Other Possibilities

This Exploration is intended to be a venture into the unknown, involving the circuit of Fig. 11.5: We have seen, in general, for various networks, that lifting elements from ground and driving the connection with a low-impedance signal, leads to various response functions, and, in particular, that a high-pass filter results if node F in Fig. 11.5 is so driven. Now, what happens here if E is lifted and driven instead, or, both E and F are lifted and driven together? Let us proceed to see!

- **Goal:**
  To revel in discovery!

- **Setup:**
  - With component values as in Fig. 11.5, and F grounded, connect a sinusoidal source of 2Vpp at 1kHz to node E.

- **Measurement:**
  a) Displaying the signal at node C, vary the frequency over a relatively wide range, and characterize the response.
  b) Repeat, with both E and F removed from ground, and driven.

- **Tabulation:**
  Input node, f, v_i, $upsilons_c$, for various input nodes and frequencies.

- **Analysis:**
  Consider what you have just learned. Explain the behaviour you find.

- **MULTIPLE-AMPLIFIER FILTERS - AN INITIAL GLIMPSE**

  In the previous two Explorations (E2.0, and E3.0), we have seen the benefit of a more sophisticated view of frequency-selective feedback around an op amp, including the possibility of positive feedback (as in Fig. 11.5 for example). Here, we will begin to look at the even greater possibilities provided by the use of two (or more) op amps. Of a number of available choices, we will chose one with an important additional property, namely its ability to simulate an inductor. Our secondary goal in this choice is to provide a degree of experience with RLC filter topologies, which are very useful at high frequencies where appropriate inductors are of a convenient size. Accordingly, we will explore the Antoniou Inductance Simulator. This circuit, a particular case of the Generalized Impedance Converter (GIC), is introduced on page 915 of the Text. The particular version we will explore is shown here in Fig. 11.6. As well, this figure provides the equivalent RLC circuit labelled with model components and corresponding grounded connections.

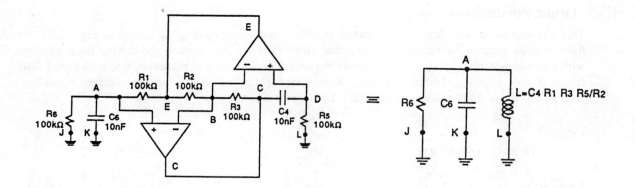

Figure 11.6 A Resonant Circuit Using an Antoniou Inductance-Simulation Circuit

E4.1 The Second-Order Bandpass Topology

- **Goal:**
  To explore the RLC bandpass filter.

- **Setup:**

  Assemble the circuit shown in Fig. 11.6, with the lower end of R_6 (node J) lifted from ground and connected to a source (node S) set to provide a 2Vpp sinewave at 1kHz. **Note**, for measurement in steps c) and d) to follow, that node B may be very sensitive to capacitive loading, such that a scope probe placed directly on it may cause oscillation. If this problem arises, use a series resistor (say 10 kΩ) to connect the probe tip to internal nodes of the circuit.

- **Measurement:**

  a) Measuring at nodes S and A, lower the frequency, noting the maximum possible output and the corresponding frequency, at f_1.

  b) At f_1, and with S as reference, measure the peak-to-peak voltage and phase of the voltages at nodes A, B, C, D, E.

  c) Vary the frequency below and above f_1, to find those frequencies (f_2, f_3) at which the gain drops (by 3dB) to 0.707 of the peak value first measured.

  d) Using S as the reference, measure the phase and amplitude of the signals at A and some of B, C, D, E, at both f_2 and f_3.

  e) Measuring peak-to-peak voltages at nodes S and A, find the frequencies (f_4, f_5) at which the response drops 20dB from its midband value.

  f) Measuring nodes S and A, raise the frequency to $2f_1, 4f_1, 10f_1, 20f_1$. At each frequency, find the ratio of available gain to that at midband, expressed directly, and in dB.

  g) Repeat step f) above for frequencies $f_1, f_1/2, f_1/4, f_1/10, f_1/20$.

- **Tabulation:**
  $f, v_s, v_a, v_b, v_c, v_d, v_e$, and phases, selectively, for various interesting frequencies.

- **Analysis:**

  Consider what you have learned about this basic RLC-like resonator. What is its center frequency, its center-frequency gain, its 3dB bandwidth and its 20dB bandwidth? What is the measured Q of the

Experiment #11–13

resonator?

- **Setup:**
    - Now, connect a second 100kΩ resistor *in series* with R_6 between nodes S and A.

- **Measurement:**
    - h) Repeat some of the previous process. In particular, find the center frequency (f_{11}), the 3dB frequencies (f_{12}, f_{13}), and the 20dB frequencies (f_{14}, f_{15}).
    - i) Repeat all of the previous steps with the second 100kΩ resistor *in parallel* with R_6.
    - j) Repeat all of the previous steps with R_6 replaced by a 1MΩ resistor.

- **Tabulation:**
    R_6, f, v_s, v_a for various values of R_6 and interesting frequencies.

- **Analysis:**
    Consider the ease with which bandwidth can be changed. Calculate the bandwidths at −3dB and −20dB, and the corresponding Q in each of the previous cases.

Measurement:
- k) It is interesting to evaluate the response to a square wave of this circuit with various values of Q. Try a 2Vpp square wave at frequencies of $f_1, 2f_1, 3f_1$, and $f_1/2, f_1/3, f_1/5, f_1/7, f_1/9$.

- **Tabulation:**
    f, v_a, wave shape, for various frequencies.

- **Analysis:**
    Consider the ability to accept and reject all or parts of a composite signal which you can observe. This is a very useful feature of RLC resonators used in oscillators, whether they use discrete or simulated inductors.

E4.2 The High-Pass Topology
- **Goal:**
    To illustrate the flexibility of filter type that the active RLC filter brings, using a high-pass example.

- **Setup:**
    - Assemble the circuit of Fig. 11.6, but with C_6 lifted from ground, and the corresponding node (K) connected to a sinusoidal source at S of 2Vpp and 10kHz.

- **Measurement:**
    - a) Measuring nodes S and A, lower the frequency, noting the change in output, noting any possible peak in output amplitude, its amplitude and frequency f_1, and the frequencies (f_2, f_3) at which the output is below the high-frequency value by 3dB and 20dB respectively. It is interesting to note the phase at these frequencies, as well.
    - b) Repeat, for another 100kΩ resistor shunting R_6.

Experiment #11–14

    c)   Repeat, for another 100kΩ resistor in series with R_6.

- **Tabulation:**
  R_6, f, v_s, v_a for three values of R_6, and various values of f.

- **Analysis:**
  Consider the degree of peaking, as a ratio and in dB, and the rate of rolloff, in the cases considered. Note how convenient it is to convert from one filter type to another!

- **Measurement:**
  d)   Try excitation with a square wave of various frequencies.

- **Analysis:**
  Consider the impact of eliminating various parts of the spectrum on the appearance of a "square wave".

E4.3 The Low-Pass Topology

- **Goal:**
  To illustrate flexibility of the active RLC filter with a low-pass example.

- **Setup:**
  ○   Assemble the circuit of Fig. 11.6, but with R_5 lifted from ground, and the corresponding node (L) connected to a sinusoidal source at S of 2Vpp and 10Hz.

- **Measurement:**
  a)   Measuring nodes S and A, raise the frequency, noting the change in output. Find the peaking frequency and gain, and the −3dB and −20dB frequencies and phases.

  b)   Repeat other aspects of E4.2 for which you have time available.

- **Tabulation:**
  f, v_s, v_a for various interesting frequencies.

- **Analysis:**
  Consider low-pass behaviour as complementary to the high-pass behaviour previously measured. What is the relationship of measured frequencies with respect to the nominal cutoff?

An easy familiarity with Op-Amp-RC Filters is very handy in for anyone working in electronics or electronic-systems design. I hope you are well on the way to this state of grace!

EXPERIMENT #12
WAVEFORM GENERATORS

I OBJECTIVES

Overall, the goal of this experiment is to familiarize you with some quite general ideas concerning the generation of waveforms using a combination of fast-acting positive feedback and delayed negative feedback, ideas which are captured in the generic term, multivibrators. For reasons both of convenience and importance in practice, we will explore circuits which employ op amps. While the op amp with its differential inputs makes some of these implementations very straightforward, the topological (connectedness) ideas are far more universal, being directly applicable to both discrete-component implementations, and logic-gate formulations, as well.

II COMPONENTS AND INSTRUMENTATION

As already noted, our amplifying device will be the 741-type op amp, available in the dual-amplifier package depicted in Fig. 12.1. Only two amplifiers are necessary, although others, if available, would allow some interesting combinations of separate circuits to be explored. As well as a number of 10 kΩ and 100 kΩ resistors, and three 10 nF capacitors, we will use eight 1N914 junction diodes, one 1N957 6.8 V zener diode, and two 10 kΩ potentiometers.

Instrumentation needed includes the usual dual power supply, a DVM, a waveform generator and dual-channel oscilloscope with ×10 probes. As well, a spectrum analyzer, if available, allows some interesting measurements to be made.

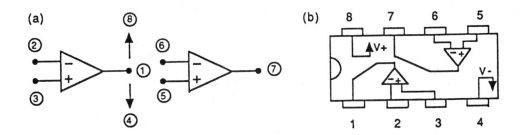

Figure 12.1 741-Type Op Amp Connections

III READING

The material on which this Experiment is based is covered primarily in Sections 12.4 through 12.9 of the Text. Of course, an understanding of basic op-amp circuits covered in Chapter 2 is assumed.

IV PREPARATION

As is the recurring pattern in this Manual, **Preparation** will be keyed to the **Explorations** to follow, by the use of the same titling and section numbering employed there, but with a P prefix.

Experiment #12-2

- ## THE SCHMITT TRIGGER, A BISTABLE MULTIVIBRATOR

P1.1 Non-Inverting Operation

(a) For the circuit of Fig. 12.2 adjusted for ±10V output and with node A grounded, sketch and label the transfer characteristic which applies from node D to node C.

(b) Modify your sketch to account for a second 100 kΩ resistor in series with an ideal diode whose cathode is at node B, all shunting R_2.

P1.2 Inverting Operation

(a) For the circuit of Fig. 12.2, adjusted for ±10V output and with node D grounded, sketch and label the transfer characteristic which applies from node A to node C.

- ## GENERATING A SQUARE-WAVE

P2.1 A Square Wave Oscillator

(a) For the circuit of Fig. 12.3 operating with output levels of ±10V, a design providing a nearly ideal triangle wave at node D is required. For what value of R_1, all other other components remaining at their present values, are the peaks of the waveform at D reduced by 1% from the value they would have if the initial charging slope had been maintained?

P2.2 Other Pulse Waveforms

(a) The circuit of Fig. 12.3, in which the output voltage is ±10V, has an external voltage V_A applied to the leftmost end of R_1. What value of V_A results in the output being positive 40% of the time?

- ## A MONOSTABLE TOPOLOGY

P3.1 Low-Frequency Operation

(a) For the circuit shown in Fig. 12.4, estimate the voltage at node E in the stable state, assuming both diode drops to be 0.7V and output levels to be ±10V. By what amount must the input at F fall to trigger the monostable?

P3.2 Higher-Frequency Operation

(a) For the circuit of Fig. 12.4, for which $v_C = \pm 10V$, sketch waveforms at nodes B and D under the conditions that a narrow 1-volt negative-going trigger pulse is introduced at node B, from node F. As the input frequency is increased from relatively low values, at what frequency does the pulse width begin to decrease? What is the slightly higher frequency at which a second trigger pulse is just missed?

- ## A MULTIPLE-WAVEFORM GENERATOR

P4.1 The Basic Square/Triangle-Waveform Generator

(a) What are the nominal limiting levels at node C, of the circuit of Fig. 12.5. What input thresholds at node A result?

(b) What frequency of oscillation do you expect? What simple change would double the frequency while maintaining the amplitude at node H? What simple change would double the frequency and halve the amplitude at node H?

P4.2 The Sine-Wave Shaper

(a) Prepare a table of sine-wave slopes as indicated in the preamble to E4.2.

(b) For v_H in Fig. 12.5 being an 8.2V peak triangle wave, with 0.7V diodes, what is the appropriate setting of P_1 for a sine-wave output at node J?

(c) For what setting of P_2 do diodes D_5 and D_7 barely conduct at the triangle peaks? For what setting does each conduct for 20°?

V EXPLORATIONS
• THE SCHMITT TRIGGER, A BISTABLE MULTIVIBRATOR

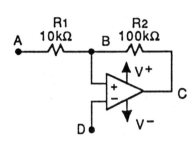

Figure 12.2 shows a basic element which recurs in the circuits to follow. It is the positive-feedback Schmitt-Trigger Bistable (Multivibrator). It is operated typically with either of nodes A or D as input, while the other is connected to a reference voltage, often ground. Because of positive feedback, the output voltage (C) is stable at one of two limiting values (a high one, L^+, and a low one, L^-) which values depend on the choice of power-supplies V^+ and V^-, and amplifier saturation characteristics. Toward the latter part of this Experiment we will consider one particular approach to making the output independent of supply-voltage and op-amp variabilities.

Figure 12.2 A Versatile Schmitt-Trigger Topology

E1.1 Non-Inverting Operation

• **Goal:**

To explore the operation of a basic noninverting Schmitt Trigger, or comparator with hysteresis.

• **Setup:**

○ Assemble the circuit of Fig. 12.2, with node D grounded and A connected to a waveform generator. Adjust the power supplies to about ±12 V. Externally trigger your oscilloscope from the generator's trigger-source output.

• **Measurement:**

a) With the generator providing a 5 Vpp triangle wave at 1 kHz to node A, display the waveforms at nodes A and C, noting the limiting voltage levels at node C.

b) Adjust the supply voltages (V^+, V^-) so that the limiting voltages at node C are closely ±10 V.

{Note that while the previous step is *not* essential, and certainly not normal in commercial practice, it is still a good example of good engineering practice: it makes subsequent measurement and interpretation very convenient. However, be quite aware that this is a measurement artifice only; Thus, in normal designs, the cost of special supply voltages makes the idea impractical, and, as we shall see, other means would be used to establish convenient symmetric output-voltage values.}

c) Displaying A and C, then A and B, note carefully the voltage values at which interesting waveform changes occur. Estimate the rise and fall times of the signals at C and B.

d) To improve your sense of the (complementary) roles of each of the resistors R_1, R_2, shunt each in turn by ones of equal value, observing the changes in thresholds at node A which result.

• **Tabulation:**

R_1, R_2, v_A, v_B, v_C, V^+_{At}, V^-_{At}, t_r, t_f, for various interesting events and three combinations of R_1, R_2.

- **Analysis:**

    Consider the operation of this circuit by sketching its transfer characteristic (v_C versus v_A). Note the hysteresis region, its width, and the simple relationship it bears to the limiting voltages (L^+, L^-) and the resistors (R_1, R_2). Note also that the symmetry of the transfer characteristic depends on the choice of $v_D = 0$. Consider the effect of raising v_D to $+2$ V (for example).

E1.2 Inverting Operation

- **Goal:**

    To explore the operation of a basic inverting Schmitt Trigger.

- **Setup:**

    O   Assemble the circuit of Fig. 12.2 with node A grounded and D connected to a waveform generator. Adjust the supplies initially to ± 12 V.

- **Measurement:**

    a) With the generator providing a 3 Vpp triangle wave at 1 kHz to node D, and observing node C, adjust the supply voltages (V^+, V^-) so that the limiting voltages at node C are closely ± 10 V.

    b) With your scope triggered directly from a fixed-voltage generator output, and displaying both nodes D and C, adjust the amplitude of the triangle wave input, so as to identify the input triggering levels, the relative timing of the output, and the minimum input signal for which node C reverses state.

    c) Now, shunt R_1 by an additional 10 kΩ resistor. First measuring nodes D and C and then D and B, identify input triggering levels and the input signal amplitude below which operation ceases.

- **Tabulation:**

    R_1, v_D, v_C, V_{Dt}^+, V_{Dt}^-, $v_{in\,min}$.

- **Analysis:**

    Consider the operation of the circuit in relationship to the version explored earlier in E1.1, noting the relative differences in polarity, input resistance, and the simplicity with which thresholds can be calculated. As we shall see, next, the signal-inverting property can be of special importance.

- **GENERATING A SQUARE WAVE**

    In general, one approach to creating an oscillator is to use a positive-feedback-based bistable element with delayed negative feedback. From a linear-circuits point of view, the idea is that of a negative feedback loop which is unstable because of the infinite loop gain which the positive-feedback element can provide. From a more digital point of view, the idea is to derive a signal from the output of the bistable which, fed back to the input after a delay, reverses the original state. A simple implementation of this idea is shown in Fig. 12.3. Here, the connections of the amplifier with R_1, R_2 form an inverting bistable whose output and input signals are of reversed polarity with input going high (or low) (beyond the

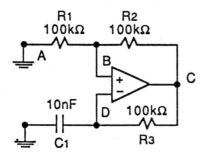

corresponding threshold) causing the output to go low (or high), respectively. Components R_3 and C_1 form a simple non-inverting delay element, whose output D follows its input C after a delay related to the time constant $R_3 C_1$. The combination is a circuit which reverses its state periodically, forming a square-wave oscillator.

Figure 12.3 A Square-Wave Oscillator, or Astable Multivibrator

E2.1 A Square-Wave Oscillator

- **Goal:**

    To explore the operation of the classical astable multivibrator.

- **Setup:**

    O Assemble the circuit shown in Fig. 12.3. To make the interpretation of your measurements somewhat easier, you may find it useful to adjust the supply voltages so that the signal levels at node C are ±10 V.

- **Measurement:**

    a) Connect your scope probes to nodes C and D, noting the relative voltage levels, time intervals, and frequency.

    b) Observe nodes D and B (with "normalized" oscilloscope channel[1]) in order to verify that the output state reverses when the capacitor voltage just reaches one of the bistable thresholds.

    c) While observing nodes C and D, shunt C_1 with a capacitor of equal value, noting the changes in time intervals and frequency, but not of waveform.

    d) With the additional capacitor removed, shunt R_1 by a 10 kΩ resistor, noting changes in waveshape at node D and the corresponding frequency.

- **Tabulation:**

    $C_1, R_1, v_B^+, v_B^-, v_C^+, v_C^-, \text{uspilon}_D^+, v_D^-, v_{Dt}^+, v_{Dt}^-$.

- **Analysis:**

    Consider the ease with which a square wave can be generated using this idea. Note that when the bistable switching threshold is made quite small, a reasonable triangle wave is also available at node D. A low-impedance triangle wave of amplitude equal to that of the square wave can be created using a second op-amp connected as an amplifying buffer with virtually the same values of R_1 and R_2 as employed in the oscillator.

---

[1] See Experiment #0.

E2.2 Other Pulse Waveforms

- **Goal:**

    To explore variants in the basic waveform.

- **Setup:**

    ○   Return to the basic circuit shown in Fig. 12.3.

- **Measurement:**

    a)   Shunt node B with a diode to ground, noting the waveforms at nodes C and D.

    b)   Reverse the diode connection and note the changes. Remove the diode.

    c)   Remove the connection of R_1 (at node A) from ground, and join it to the tap of a 10 kΩ potentiometer R_p whose ends are connected to the $\pm$ supplies. Note the effect of varying the potentiometer on the waveforms at C and D.

    d)   Now, with a DVM to measure the dc value of the voltage V node A, and an oscilloscope at node C to measure the widths (at the zero-volt level) of the positive and negative output signals (T^+ and T^- respectively), take enough measurements to characterize the relationships between control voltage and pulse widths, duty cycle and frequency.

- **Tabulation:**

    v_B^+, v_B^-, v_C^+, v_C^-, v_D^+, v_D^-; V_A, T^+, T^-.

- **Analysis:**

    Consider the various flexibilities in pulse-width control which the circuit presents. What do you think will happen if a series combination of a diode and a 1 kΩ resistor is used to shunt the resistor R_3? (Try it!)

• A MONOSTABLE TOPOLOGY

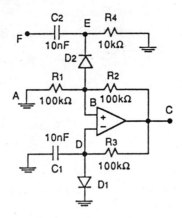

Figure 12.4 A Monostable Multivibrator

In the circuit of Fig. 12.3, if a diode is connected from node D to ground (say with cathode on ground), node D is clamped at +0.7 V or so, and thereby prevented from reaching the bistable's upper threshold. As a result, circuit operation stops with node C at its high value (say +10 V) and D at 0.7 V. Thus the circuit is monostable. To put it in the active region ever again requires some additional external input.

Such an arrangement is shown in Fig. 12.4, where D_1 is the clamping diode just described. Components D_2, C_2 and R_4 allow a signal at node F to control the upper threshold (the voltage at node B) of the bistable.

E3.1 Low-Frequency Operation

- **Goal:**

    To explore basic one-shot operation.

- **Setup:**
    - ○ Connect the circuit as shown in Fig. 12.4, with a square wave of 2 Vpp amplitude at 100 Hz applied to node F.

- **Measurement:**
    - a) Display nodes F and C on your oscilloscope, noting the relative timing. Adjust the supplies to establish the limits of the output signal at node C at ± 10 V.
    - b) With node C as a reference, examine nodes D, B, E, F in turn, noting first the effect and then its cause (as an exploratory reversal of the usual cause-effect analysis). Note the relative voltage levels and timing of all the waveforms.
    - c) Displaying the signals at nodes F and B, lower the square-wave input amplitude until normal operation just ceases. Note the minimum signal required.
    - d) At an input voltage barely above the triggering threshold, examine nodes E, B, D in turn, carefully noting the activity just at the triggering point. Adjusting the input amplitude intermittently around the threshold, is likely to aid in your understanding. Use small sketches to record your observations.
    - e) Connect a large capacitor (0.1 µF to 1 µF) from node C to ground and repeat the previous two steps. {The capacitor will slow the bistable response enough to allow you to see the separated effects of input and feedback signals.} Prepare a sketch of the activity of nodes B and D.

- **Analysis:**

    Consider normal operation at relatively low repetition rates: the range of triggering signal amplitudes, the roles of D_1 and D_2, the effective time constants at nodes D and E, and the role of C_2.

    Note that a generally useful approach to gaining understanding of the roles of each of the components in a circuit such as this is to shunt each of them, in turn, with one of equal value, while examining appropriate nodes under appropriate conditions.

    - f) For example, while displaying nodes B and E, shunt C_2 and R_4 in turn, with input square-wave amplitude adjusted just above the threshold.

E3.2 Higher-Frequency Operation

- **Goal:**
    To examine the one-shot operation of high-rate triggering signals.

- **Setup:**
    - ○ Use the circuit of Fig. 12.4, with an input signal of about twice the minimum amplitude.

- **Measurement:**
    - a) While you examine nodes B and D, raise the input frequency slowly, until the mode of operation changes. Note the critical frequency, f_1.
    - b) With a frequency of about $1.1 f_1$, raise the input amplitude until operation becomes more normal, noting the overall effects and the length of the output pulse which results.
    - c) With the amplitude a few percent higher than that just found, raise the frequency to around $2f_1$, then $3f_1$, etc.

- **Tabulation:**

    v_F, f_{in}, f_{out}, where f_{in} is the trigger input frequency and f_{out} is the frequency of pulses at C.

- **Analysis:**

    Consider the modes of operation just demonstrated, and the corresponding limitation on triggering frequency. How are f_1 and pulse period T related? These restrictions are a result of the charge-recovery process on C_1, following each output pulse. Explain? If there is a need to retrigger a one-shot more often then every few pulse periods, other more sophisticated designs may be necessary. The following exploration step indicates one simple direction for improvement.

- **Setup:**

    ○   Augment the circuit of Fig. 12.4 with a diode D_3 and resistor $R_5 = 10$ kΩ in series, connected from node C to D with the diode cathode on node D.

- **Measurement:**

    a)   Repeat some of the measurements above, particularly for frequencies f_1 and above. Examine the waveform at node D.

- **Analysis:**

    Consider the improved recovery time that R_5 produces. What does the maximum trigger rate for normal operation become?

- ## A MULTIPLE-WAVEFORM GENERATOR

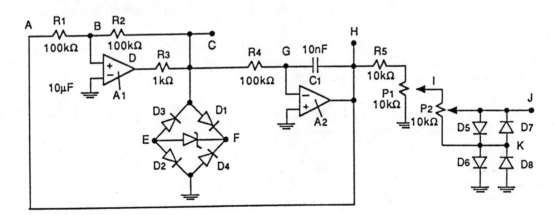

Figure 12.5 A Function Generator

The circuit shown in Fig. 12.5 is a simplified version of the general-purpose laboratory instrument called a Function Generator, or Waveform Generator that you use in most experiments. It follows the basic principle enunciated earlier in conjunction with Fig. 12.3: that of a bistable connected in a negative-feedback loop with a delay element.

Here A_1 with R_1, R_2, R_3, D_1, D_2, D_3, D_4 and Z, constitute a non-inverting bistable (Schmitt Trigger) circuit having a shunt regulator at its output to ensure power-supply-voltage independence. The diode bridge (D_1, D_2, D_3, D_4) allows a single zener diode, Z, to function as a shunt regulator for both positive and negative output voltages at node C. Resistor R_3 limits the zener-diode and amplifier-output currents. Amplifier A_2 with R_4 and C_1, comprise an inverting integrator which provides a delay in the signal fed back to the bistable input. The circuit consisting of R_5, P_1 and P_2 with diodes D_5, D_6, D_7, D_8 is a very basic 5-segment wave shaper capable of providing an approximate sine-wave output at node J.

E4.1 The Basic Square/Triangle-Wave Generator

- **Goal:**

    To explore the operation of the basic function generator as an important example of a small electronic system.

- **Setup:**

    ○ Assemble (at least) the part of Fig. 12.5 that includes A_1, A_2 and the connection from node H to node A. Here Z is a 1N967 6.8 V zener diode. Use ±15 V supplies.

- **Measurement:**

    a) Display the signals at nodes C and H. Note the two waveforms, their peak values, and frequency, f_2.

    b) In turn, shunt R_1, R_4, C_1 by components of equal value, and note the effects on signal amplitudes and frequencies, f_3, f_4, f_5.

    c) With the circuit in its original form, display the waveform at node C with those at nodes G, H, B, D, E, F, in turn, as the basis of a structured timing sketch.

    d) While displaying the waveforms at nodes C and H, short out the zener Z intermittently, noting the changes in amplitude and frequency.

    e) While displaying nodes C and H, remove zener Z, observing the overall effect. Note that without Z, operation depends on the relative saturation voltages of A_1 and A_2. Operation is guaranteed only for R_1 less than R_2 by some (small) amount. (Why?)

- **Analysis:**

    Consider the variety of waveforms available, and the means for their control. Prepare an organized well-labelled timing sketch, including at least nodes A, B, C, H.

E4.2 The Sine-Wave Shaper

If not already done, assemble the rightmost part of Fig. 12.5, the sine-wave shaper, initially with potentiometers P_1 and P_2 in the middle of their ranges. Your challenge now is to adjust these potentiometers to create the most satisfactory pseudo-sine wave at node J from the triangle wave at node H. In general, the strategy involves realizing a) that P_1 can be used to control the slope of the output for small outputs in the range ±0.4 V or so, b) that the peak voltage is limited to ±1.4 V or so, and c) that P_2 controls the region in between. There are a great many possible approaches to adjustment, depending on available instruments: whether you have a gated (or triggerable) waveform generator, a distortion analyzer, a spectrum analyzer, etc.

- **Goal:**

    To explore the operation of a triangle-wave-to-sine-wave converter.

- **Setup:**

    ○ We will assume the worst: that your Laboratory is equipped, like the majority of academic laboratories, sparsely! Thus we will implement a waveform-matching process. Accordingly, as a preliminary move to obtain a reference waveform, connect your oscilloscope to the commercial waveform generator with which you are supplied.

Experiment #12–10

- **Measurement:**

    a) Adjust the generator output and oscilloscope triggering to provide a stable sine-wave display of peak amplitude 1.4 V and frequency f_2 (as measured in E4.1, first step).

    b) Display a complete cycle with the rising-edge zero-crossing nicely aligned with your display graticule, with the display set at 0.5 V per major division.

    c) Lay a piece of thin paper over your screen and trace the waveform, noting in particular its initial slope in the first 0.2 V or so, its slope value at ±0.5 V, at ±1 V, and at the 45° points. You are now ready to proceed to the actual process of adjusting the sine-wave shaper!

- **Setup:**

    O   Return to the circuit of Fig. 12.5.

- **Measurement:**

    d) Observing nodes H and J, verify that the frequency of your circuit is f_2 as you used in the sketch described above. Otherwise, if a horizontal display vernier-adjustment option exists on your oscilloscope, adjust it to compensate, or, else, redo your sketch.

    e) Likewise, if the peak amplitude of the signal at node J is not as in your sketch, compensate by adjusting the variable vertical-gain control of the oscilloscope.

    f) With your sketch overlayed on your scope face and the display modified appropriately, adjust P_1 in your circuit until, the slope of the waveform at node J around zero volts matches that of your sketch.

    g) Adjust P_2 to make the output waveform fit your sketch as well as you can.

    h) Remove your sketch from the scope face and revel in the beauty of the sinewave you have created. Make some careful measurements of your waveform for comparison against actual sine values.

    i) If a spectrum analyzer is available, connect it to the output of your circuit, and measure the harmonic content of your waveform.

    j) Make very small changes in the settings of P_1 and P_2 in an attempt to reduce the total harmonic content of your pseudo-sine wave.

- **Analysis:**

    Consider the virtue of your output waveform by various comparisons with a real sine wave, using several different approaches: your aesthetic judgement, a large-scale graph, or harmonic-content measurement.

The circuits explored here are very handy for the generation of signals at modest frequencies in real-life instrumentation and testing applications. We all hope they have already become second nature to you, ready "at the drop of a hat", so to speak, for any need that, may arise, such as over drinks as an ice-breaker, say, as in: "Last week, I worked on this neat circuit --------".

EXPERIMENT #13
CMOS LOGIC CHARACTERIZATION

I OBJECTIVES

The overall objective of this Experiment, is to familiarize you with the internal workings of a typical CMOS gate as seen through the terminals of a small-scale integrated (SSI) circuit package. For this purpose, an earlier logic form, non-buffered CMOS, will be stressed. While the same experiments can be performed on buffered CMOS as a enhancement of this experience, some of the results would be less informative, though no less interesting and exciting! Some diverse applications of CMOS will be encountered, as well. There are a lot of interesting things to explore here. If pressed for time, concentrate on E1.2, E1.3, E3.1, and E4.1.

II COMPONENTS AND INSTRUMENTATION

The logic-gate form on which we will concentrate is the Quad 2-input NAND gate in the 14-pin DIP, whose pin arrangement is shown in Fig. 13.1. As noted, we will concentrate on the non-buffered version, the CD4011UB, for which the individual gate circuit is shown in Fig. 13.2.

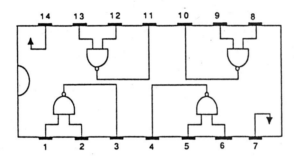

Figure 13.1 The 4011-Type Quad 2-Input NAND Package

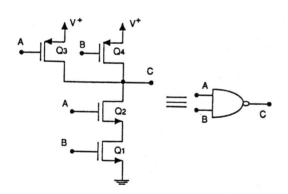

The buffered version, CD4011B, which performs the same logic function, has two tandem inverters connected to drive the output. Correspondingly, the input transistors, being buffered from the load, can be smaller, with considerably smaller input capacitances.

Two CD4011UB will be used. As well, resistors of values 100 Ω, 1 kΩ, 10 kΩ, 100 kΩ and 1 MΩ, of ordinary precision (5%, or even 10%), and capacitors of values 1.0 µF (ceramic monolithic) (2), 1000 pF, and 100 pF, are required. For instrumentation, two supplies, a DVM, a waveform generator, and a 2-channel oscilloscope with ×10 probes will suffice. A second DVM is useful, if available.

Figure 13.2 The Basic CMOS NAND Circuit

Experiment #13–2

III READING

Concentration in this Experiment will be on Sections 13.1 through 13.3 of the Text. As well, Section 5.8 will provide some useful background.

IV PREPARATION

As is the recurring pattern in this Manual, **Preparation** will be keyed to the **Explorations** to follow, by the use of the same titling and section numbering employed there, but with a P prefix, rather than an E.

• THE NAND GATE – BASIC FUNCTIONS

P1.1 The NAND Logic Function

(a) Prepare a truth table for a NAND gate, using the H/L notation, H for high and L for low.

P1.2 Measurement of Propagation Delay

(a) A ring of five CMOS gates is found to oscillate at 2 MHz. What is the average propagation delay of each of the gates?

(b) For the ring described above in (a), with one of the gate outputs shunted to ground by a 100 pF capacitor, the oscillation frequency reduces to 1.43 MHz. What is the extra delay associated with the loaded gate for each transition (be careful!)?

(c) From the data in (b) above, estimate the average (capacitor-charging) output current capability of the capacitor-loaded gate.

P1.3 Gate Threshold(s)

(a) Sketch the setup for measuring V_{th} and corresponding supply current, as described in the first step of E1.3.

(b) For the CMOS NAND circuit as shown in Fig. 13.2, with $V^+ = 5V$, input B connected to V^+, and A and C joined, find V_{th} and I_{Q1}. For all transistors, $k = 0.5$ mA/V^2, and $|V_t| = 1V$.

(c) Repeat (b) for $V^+ = 15V$.

• GRAPHICAL CHARACTERIZATION

Review your understanding of the use of an oscilloscope in presenting a display of a transfer characteristic in which V_{out} is displayed on the vertical axis, while V_{in} is presented horizontally. Make yourself familiar with all of the applicable switches and controls (See Experiment #0).

P2.1 Input, Transfer and Supply-Current Characterization

(a) For the 2-input NAND of Fig. 13.2 using a +5V supply, and devices for which $|V_t| = 1V$ and $k = 0.5$ mA/V^2, with $V_B = V^+$, and input at node A, find V_{th} (using the results of P1.3 b) above), V_{IL}, V_{IH}, together with the corresponding output voltages and supply currents. Plot these three data points with results for inputs exactly one threshold voltage from each supply line, to provide the general shape of the transfer and supply-current characteristics.

P2.2 Variation with Input Connection and Power-Supply Voltage

(a) Repeat the process described in P2.1 a) with nodes B and A interchanged, ie, with $V_A = V^+$ and the input at V_B.

• Transfer-Characteristic-Slope Evaluation

(a) When you use a 100 kΩ resistor as a probe at the end of a DVM lead in order to isolate the lead's capacitive and antenna-like properties, what reading error results? Note that R_{in} of the DVM is typically 10 MΩ. If two such resistors are used to allow a floating measurement, what does the error become?

P3.1 Maximum Gain

(a) For the NAND gate analyzed in P1.3 b), for which the Early Voltage V_A is said to be 50V, estimate the small-signal gain v_c / v_a for nodes A and C joined by a very large resistor.

P3.2 Thresholds V_{IL}, V_{IH}

(a) For the NAND gate connected as in Fig. 13.5, but with $V_B = V^+$, input at A, output at C, what are the small-signal gains v_c / v_a at $V_A = V_{IL}$ and $V_A = V_{IH}$?

- **OUTPUT-DRIVE CAPABILITY**

P4.1 Short-Circuit Output Current

(a) For the NAND gate described in P1.3 a) with $V_B = V^+$ and $V_A = 0$, what current flows when the output is short-circuited to ground? What does the current become if node B is also grounded?

(b) What does the output current become with $V_B = V^+$ and $V_A = 0$, when V_C is held at $(V^+/2)$?

P4.2 Dynamic Output Characterization

(a) Using the data calculated in P4.1 a) b), sketch the output voltage v_C as it rises from 0V to 2.5V (on its way to 5V), driving a 1000 pF load. What happens if the input is connected not simply to input A, but to both inputs?

(b) Sketch the Schmitt Trigger arrangement described in the early part of E4.2. involving inverter-connected gates N_1, N_2, N_3, and a 10 kΩ resistor. Notice that when the output of N_3 is high, the input of N_1 must go above its normal threshold by enough to supply the current needed by the 10 kΩ resistor at the regular threshold of N_2 before N_2 and N_3 will switch. Similarly, for the output of N_3 low, the input of N_1 must go below its normal threshold for switching. Notice that once the output of N_3 begins to change, the current in the 10 kΩ reduces and the output of N_1, whose load is being removed, changes by itself (even if the input of N_1 moved no further). This interesting regenerative behaviour is what characterizes a Schmitt Trigger and accounts for its fast output transitions.

(c) For gates, connected as described in the first section of E4.2, and for which the usual V_{th} is at 2.5V, and $|k|$ is approximated to be 1.0 mA/V² for the combined set of four transistors in the output of N_1, estimate (roughly) the (modified) low- and high-threshold voltages at the input of N_1.

V EXPLORATION

- **THE NAND GATE – BASIC FUNCTIONS**

The internal arrangement and symbol for the basic NAND gate are shown in Fig. 13.2. We will normally use a supply voltage V^+ of +5 V, although operation from 3 V to 18 V is possible: *Be Very Careful* at the high-voltage end, not to exceed the 18 V rating, (though in practice rather more can be tolerated). Note that an open input to a CMOS gate represents an undefined logic state, but, more seriously, one in which the current from the supply to ground through the gate transistors can be quite large. Accordingly, *unless otherwise specified*, ground one input of each NAND gate (say pins 2, 5, 9, 12).

E1.1 The NAND Logic Function

- **Goal:**

    To verify the logic function of one of the four gates (the one having pins 1,2,3) in the Quad NAND package.

- **Setup:**

    ○ Install the 4011 IC on your prototyping board with + 5 V connected to pin 14 and ground to pin 13. Referring to Fig. 13.1 and Fig. 13.2, connect together pins 1, 6, 8, 13 as input A, pins 2, 5, 9, 12, as input B. Pins 3, 4, 10, 11, *not joined*, are *separate* outputs C.

- **Measurement:**

    a) Apply all possible combinations of ground and +5V connections (4 of them) to inputs A and B, noting the corresponding output voltages at the four separate outputs [nodes C (pins 3, 4, 10, 11)]. Prepare a truth table with columns A, B, C and entries of 0 and 5 (volts).

- **Analysis:**

    Consider the logic function you observe. What single input combination is needed to lower the NAND output to 0 V?

E1.2 Measurement of Propagation Delay

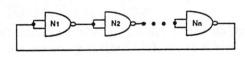

A ring of n inverting logic gates connected as shown in Fig. 13.3 oscillates for $n \geq 3$ (n odd) (always, in practice). For $n = 1$, the output is normally dc-stable at the switching threshold, although oscillation is possible for complex gates. For $n \geq 2$ (n even), the circuit is bistable, that is alternating gate outputs around the loop (of 4, 6, etc) are either high (H) or low (L), in a stable state.

Figure 13.3 A Logic Ring

[**Note** that actually, an even-numbered ring can be considered to be an oscillator as well, but one with a frequency of oscillation at 0 Hz, for which, at any moment, at any node, we are looking at only one half cycle of the oscillation, which half cycle will end in infinite time!]

- **Goal:**

    To explore the behaviour of the smallest astable logic ring.

- **Setup:**

    O   Connect three of the NAND gates, with inputs joined, in a ring of 3, with the 4th NAND disabled by having 1 input grounded.[1] Using ×10 probes, connect one channel of your oscilloscope to 1 of the gate outputs and the other to the next output along the ring.

- **Measurement:**

    a) Note the waveform shape, peak voltages, and frequency.

    b) Note the relative delays at the rising and falling edges of the waveforms.

    c) Remove one probe and note any change at the other node.

    d) Remove probes from the ring, and connect *one input* of the spare gate to one of the ring outputs, with the other NAND input connected to 5 V. With one probe connected to the output of the 4th NAND, measure peak-to-peak voltage and frequency.

- **Tabulation:**

    $f_1, f_2, f_3, v_p^+, v_p^-$.

---

[1] Note that it is important that unused CMOS gates be disabled appropriately (with a grounded input for a NAND, and a high input for a NOR) in order to prevent them from entering the high-current state.

Experiment #13–5

- **Analysis:**

    Consider the case with which you can measure average propagation delay. What is its value for these gates? Note the impact of loading on performance.

- **Setup:**
    - ○ For interest, in some of the following steps, insert your DMM (connected as a DCM) to measure current in the supply lead (pin 14) of your IC. Be certain that you have an appropriate low-inductance capacitor (1 µF) connected quite directly from pin 14 to pin 7 of your IC. Ensure that any spare gate is disabled (with one input grounded).

- **Measurement:**
    - e) With the basic ring of 3, and one probe connected, raise the supply voltage from 5 V, to 10 V, and then to 15 V, noting the new frequencies and peak signal values.
    - f) Now lower the voltage supply below 5 V. What is the voltage at which operation ceases? At what voltage does it start again? Explain?

- **Tabulation:**

    V^+, f, v_p^+, v_p^-.

- **Setup:**
    - ○ Using a second IC and a 5 V supply, construct a larger ring of 5 inverting gates with the three surplus gates, each having 1 input connected high, have the other connected to a ring output, each separated (in two cases) by an intervening ring gate.

- **Measurement:**
    - g) With one of these buffer outputs as reference, examine the other two, noting their improved waveforms, their peak values, relative delays and, of course, the frequency.
    - h) While measuring two of the in-phase buffered outputs, load the intervening node with a 100pF capacitor to ground, noting the effect on relative delay, transition times, and overall frequency.
    - i) Now for the 5-element setup, repeat the supply-voltage Exploration, measuring at least the frequency, for a supply voltage of 10 V, and of 15 V.

- **Tabulation:**

    C, t_{PHL13}, t_{PLH13}, for $C = 0$ or 100 pF, and also V^+, f, v_p^+, v_p^-.

- **Analysis:**

    Consider a set of interesting properties of the ring oscillator, including the dramatic effect of supply voltage on oscillation frequency. Use the result of loading by a 100 pF capacitor to estimate the average gate drive current, and the average ring-node capacitance.

E1.3 Gate Threshold(s)

- **Goal:**

    In this Exploration, our goal is to examine the threshold voltage, V_{th}, that input voltage at which the input and output voltages are equal. This measurement is particularly easily done for an inverting logic gate, simply by joining the input and output leads to force input-output voltage equality. The result is a negative-feedback amplifier having a lowered output resistance, and with output voltage stabilized at V_{th}. The loop is generally stable for simple logic gates. For multistage Buffered CMOS gates this may not be the case due to the multiple (but staggered) poles involved. However, with CMOS, stability is easily established by introducing a dominant pole using a large-valued resistor in the feedback connection, with a large capacitor from the gate input to ground. Though this idea is quite general, it is especially convenient for CMOS since the gate current is infinitesmal and a very large resistance can be used.

    For a 2-input CMOS NAND, there are three possible connections to be evaluated, two with one input high and the other connected to the output, and one with both inputs connected together to the output.

- **Setup:**

    ○ For one gate (of the four in your IC), connect both inputs to the output. Ground at least one input of each of the other gates. Connect pin 14 of the IC to +5 V via a second DVM connected to measure current. Since the meter leads are quite inductive, ensure that there is a low-inductance bypass capacitor (1.0µF) connected directly between the IC power pins with short leads.

- **Measurement:**

    a) Use your primary DVM to measure the (joined) input (threshold) voltage, and note the supply current if a second DVM is available.

    b) Connect a 1kΩ resistor from the output, first to ground, and then to +5 V while measuring the input/output voltage.

    c) With the resistor removed, raise the supply voltage, first to 10 V, then to 15 V, measuring the input voltage (and supply current) which result.

    d) With the supply again at 5 V, connect one gate input to the positive supply while the other remains connected to the output. Measure the threshold voltage (and the supply current which results). If time permits, repeat this for the supply raised to 10 V and 15 V.

    e) Repeat for the other input combination.

- **Tabulation:**

    V^+, V_A, V_B, V_C, I_{supply}, for three-input combinations, and three supply voltages.

- **Analysis:**

    Consider what you have learned about the threshold voltage. For a 5 V supply, rank the various thresholds from low to high. What is the range? Can you deduce which of the inputs is associated with the grounded-source NMOS transistor? What can you say about the effect of power-supply variation on threshold voltage? On supply current? Use your measurements to quantify these relationships. Do you have enough data to find V_H and k for all the transistors?

- **GRAPHICAL CHARACTERIZATION**

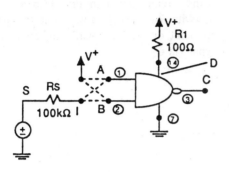

Figure 13.4 A NAND Test Circuit

The circuit we will examine in this Exploration is depicted in Fig. 13.4, where only the gate under test (GUT) is shown. For each of the other three gates, at least one input should be grounded. Resistor R_1, connected in the common gate-supply lead, is to be used to display transition currents. Resistor R_S limits the input current when the input voltage exceeds that of the input-protection circuits. Use your oscilloscope in the $X - Y$ display mode (See Experiment #0) with node S (the triangle-wave generator) connected as the horizontal input.

E2.1 Input, Transfer and Supply-Current Characterization

- **Goal:**

  To use the input sweeping technique to characterize an individual logic gate.

- **Setup:**

  ○ With a 5 V supply voltage and 12 Vpp triangle-wave input at 1 kHz, connect inputs A and B together (pins 1, 2) to node I,

- **Measurement:**

  a) Display voltages at nodes A and C in turn. Sketch and label the displays. From the combination, estimate V_{th}, V_{IL}, V_{IH} and the maximum transfer slope. From the former characterize the input-protection network.

  b) Now, for the same conditions as above, display nodes C and D, the latter with high gain and ac coupling selected.

- **Analysis:**

  Consider the convenience of this technique for displaying and quantifying transfer, input, and supply characteristics.

- **Measurement:**

  c) With both inputs joined to the triangle-wave source, and node C displayed, adjust the input source for maximum amplitude. Then, raise the supply voltage, first to +10 V, then to 15 V. Note that the available input waveform may not be large enough to produce a complete transfer characteristic at high supply voltages. Identify corresponding values of V_{th}.

  d) Returning the supply to 5 V, remove first one and then the other gate input from the input source connection (node I), and connect it to the positive supply. Try to do this relatively quickly so as to facilitate a direct comparison in your mind between pairs of transfer characteristics. Identify three values of V_{th}.

- **Analysis:**

  Consider a summary of the effects of input and supply-voltage variation by tabulating observed values of the supply voltage and V_{th}, the input voltage for which input and output voltages are equal. You will see another way to measure the transfer-characteristic slope in the Exploration to follow.

Experiment #13–8

- **TRANSFER-CHARACTERISTIC-SLOPE EVALUATION**

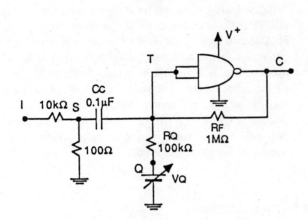

In the circuit shown in Fig. 13.5 a second supply, V_Q (with R_Q and R_F) allows one to vary the operating point of the gate while injecting a small input signal via C_C.

Figure 13.5 Measuring the Transfer-Characteristic Slope

Note: To reduce the effect of DVM leads on signal conditions, when measuring dc voltages in an active circuit, use a series resistor (say 100 kΩ) at the DVM probe tip.

E3.1 Maximum Gain

- **Goal:**

  To evaluate the maximum slope of the transfer characteristic.

- **Setup:**

  O Assemble the circuit shown in Fig. 13.5 with a 2 Vpp sine wave at 10 kHz connected to node I, $V+ = 5.0$ V and nodes T and C displayed on your oscilloscope.

- **Measurement:**

  a) Adjust V_Q to maximize the peak-to-peak voltage at C. Measure the pp voltages at T, C and the dc voltages at T, Q, C. Find the gain from T to C.

  b) Connect a 100 kΩ resistor, R_X, from node C to node Q, and find the gain from T to C.

  c) With R_X removed, adjust V_Q to find two operating points at in which the gain from T to C is reduced by (say) 10% from its peak value. Measure the corresponding dc voltages at Q, C.

- **Tabulation:**

  R_X, V_T, V_Q, V_C, v_t, v_c, for $R_{X1} = $ 100 kΩ or ∞, for v_c as large as possible or 90% of that value.

- **Analysis:**

  Consider the significance of the point of maximum gain. What is the maximum gain value? From your results with R_X, what can you conclude about the corresponding g_m and output resistances of the output transistors.

E3.2 Thresholds V_{IL}, V_{IH}

- **Goal:**

    To directly measure gate threshold.

- **Setup:**

    ○   Use the setup as in Fig. 13.5 and E3.1, with no additional components connected.

- **Measurement:**

    a) While measuring peak-to-peak voltages at nodes T and C, lower V_Q until the gain from T to C is -1 V/V. Then, measure the DC voltages at T, Q, C.

    b) Repeat the process above by raising V_Q, until the gain is -1 V/V once again.

- **Tabulation:**

    V_T, V_Q, V_C, v_t, v_c for $v_c / v_t = -1$ V/V.

- **Analysis:**

    Consider the ease with which you have directly identified V_{IL} and V_{IH}.

- **Measurement:**

    c) As a means by which to visualize the rate of change of slope of the transfer characteristic near V_{IL}, lower V_Q until the gain from T to C is $-\frac{1}{2}$ V/V. Then raise it until the gain becomes -2 V/V. Note the dc voltages at nodes T and C in each case.

    d) Repeat some of the above, with one or the other of the two NAND inputs connected to the supply V^+.

    e) Repeat some of the above, with the supply voltage raised to $+10$ V or to $+15$ V, if time permits.

- **Tabulation:**

    V^+, V_T, V_Q, V_C, v_t, v_c, for $V^+ = 5, 10, 15$ V, and $v_c / v_t = -\frac{1}{2}, -1, -2$.

- **Analysis:**

    Consider using some of the gain measurements, and the input and output bias voltages at which they are found to calculate k and V_t values for the p-and n-channel gate transistors.

- **OUTPUT-DRIVE CAPABILITY**

    In this Exploration we will investigate two approaches to characterizing the drive capability of a CMOS gate: The first is a direct static measurement. The second is a dynamic one involving charge and discharge of a load capacitance.

E4.1 Short-Circuit Output Current

- **Goal:**

    To make a direct measurement of gate short-circuit current.

- **Setup:**

    ○   Prepare the gate to be tested by joining its inputs together and connecting them to its output by a small resistor (say 100 Ω). Connect your DVM across this resistor. Initially, set the supply to 5.0 V.

- **Measurement:**

    a) Short *the input* to ground, and measure the drop across the resistor, and, thereby, the short-circuit outward-directed output-stage current.

    b) Short *the input* to the positive supply, measuring the resistor drop and the corresponding inward-directed output-stage current.

    c) Repeat steps a) and b) above for supply voltages of 10 V and 15 V if time permits. **Note:** Particularly in the latter case, use intermittent short circuits to minimize device heating.

- **Tabulation:**

    V^+, v_I, v_O, I for $v_I = 0$ V or 5 V and $V^+ = 5$ V, 10 V, 15 V.

- **Analysis:**

    Consider what you can learn from these measurements (or even a subset of them) about $|V_t|$ and k of the output devices.

E4.2 Dynamic Output Characterization

In this exploration, our need is for a driving waveform which is relatively ideal, that is, one whose transitions are very rapid in comparison with those expected at the measured output. In a very-well-equipped laboratory, this need may be best met by a very high-performance pulse generator. Here, our intent is to explore an alternative, that of improving the transition times of an available generator using other logic gates, first simply as amplifiers, and then in the positive-feedback structure called a Schmitt Trigger.

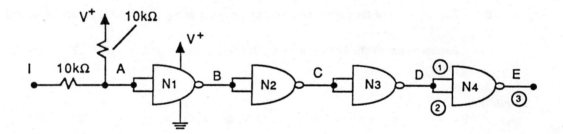

Figure 13.6 A NAND-Gate Cascade

- **Goal:**

    The ultimate goal is to be able to characterize the dynamic load-driving capability of a CMOS gate. In order to facilitate this, a test setup will be prepared and the implementation and the impact of a Schmitt trigger demonstrated.

- **Setup:**

    ○ Assemble the circuit shown in Fig. 13.6, where N_4 is the gate to be tested, and one or more of N_1 through N_3 can be perceived as amplifiers for improving the transition times at node D. The input is driven initially by a 10 Vpp square wave at 100 kHz. Here, the input resistors serve both to translate the symmetric generator output to the gate operating range, and to avoid over-driving the input-protection diodes.

Experiment #13–11

- **Measurement:**

    a) Display the waveforms at nodes B and D on your 2-channel scope using probes. What are the transition times for each?

    b) Repeat the previous step with the input switched to provide a triangle wave at I.

    c) Connect a 10 kΩ resistor, R_F, between nodes B and D and note the change in the waveforms at nodes B and D.

    d) Measuring nodes B and C, intermittently remove R_F, noting waveform changes.

- **Tabulation:**

    R_F, v_B, v_C, v_D, t_{tLHD}, t_{tHLD}, t_{tLHB}, t_{tHLB}, [waveform sketches].

- **Analysis:**

    Consider the benefit of positive feedback via R_F. What is happening? {The circuit formed by N_2, N_3, with the output circuit of N_1 is a particular form of *Schmitt Trigger*.}

- **Setup:**

    ○ Using the circuit of Fig. 13.6, with a 10 Vpp square wave at I (and R_F installed).

- **Measurement:**

    e) With a probe connected at node E, measure the rise and fall times.

    f) Intermittently connect a 1000pF capacitor (C_L) between node E and ground, noting detailed changes in waveform.

    g) With C_L connected, measure the times at which the transition is 10%, 20%, 50%, 80%, 90% complete.

- **Tabulation:**

    C_L, $t_{0.1}$, $t_{0.2}$, $t_{0.5}$, $t_{0.8}$, $t_{0.9}$, both for rising and falling transitions.

- **Analysis:**

    Consider what you can learn from the waveform of a capacitively-loaded output. For each direction of transition, estimate the short-circuit output currents. What are the currents halfway through the swing? At 90% of the way? What can you say about V_t, k? What you can say about the output-stage driving capability of a CMOS NAND?

CMOS is a very important logic structure in very-large-scale integrated (VLSI) systems where power dissipation is important. Such CMOS chips have been built with many millions of transistors! In fact, CMOS chips are responsible for the latest wave of microprocessor developments.

NOTES

EXPERIMENT #14
TTL LOGIC CHARACTERIZATION

I OBJECTIVES

The objectives of this Experiment are to familiarize you with techniques for evaluating a TTL logic gate as a circuit-element. Concern will be first for measurement of circuit properties which affect use as a system element. Second, there will be some effort spent on identifying and characterizing some important TTL structures internal to your IC.

The particular concentration will be on Low-Power Schottky TTL as a popular low-cost relatively-high-performance TTL variant. The techniques illustrated are generally applicable to other members of the TTL family including Advanced Low-Power Schottky TTL. You will note that this Experiment is quite long, too long to finish in one regular laboratory session. If pressed for time you should concentrate on Sections E1.2, E1.3, E2.1, and E3.1.

II COMPONENTS AND INSTRUMENTATION

For various reasons, including convenience, availability, flexibility and typicality, we will employ the quad two-input NAND gate package (a 14-pin DIP) whose pin diagram is reproduced here as Fig.14.1. For Low-Power Schottky TTL, the generic part number is 74LS00. Though only a single IC is essential for most of the explorations, two will provide some optional experimental flexibility, a minor glimpse at component-to-component variability, and an additional possibility of component substitution during troubleshooting. As well, a few resistors and capacitors are needed: two 100 Ω, two 1 kΩ, one 470 Ω, and one 10 kΩ, along with two 100 µF tantalum and two 1 µF ceramic capacitors, one of each of the latter two being reserved for power-supply bypassing. Instruments required are the usual ones: a DVM, two power supplies, a waveform generator, and a dual-channel oscilloscope with ×10 probes. The oscillosscope bandwidth should exceed 50 MHz if possible. A frequency counter and fast pulse generator are useful, but not essential. A second DVM would also be helpful.

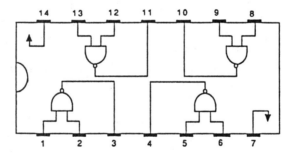

Figure 14.1 The Quad NAND Package

III READING

Appropriate background for this Experiment can be found in Sections 14.3 through 14.5 of the Text.

IV PREPARATION

As is the recurring pattern in this Manual, **Preparation** will be keyed to the **Explorations** to follow, by the use of the same titling and section numbering employed there, but with a P prefix.

Experiment #14–2

- **LOW-POWER SCHOTTKY (LS) TTL – THE BASICS**

P1.1 The NAND Logic Function

(a) For the LS TTL circuit shown in fig. 14.2, what would you expect the voltage at the open input diodes to be?

(b) Prepare a truth table for the NAND logic function, using the $H-L$ notation referred to in E1.1.

P1.2 Operating Speed – Propagation Delay

(a) Using data provided on page 1195 of the Text, estimate the oscillation frequency of rings of 3, 5, and 7 LS TTL gates.

P1.3 Switching Threshold, Transfer Characteristics, Input Characteristics

(a) In the circuit of Fig. 14.3, using a gate for which $V_{th} = 1.3V$ and $I_{IL} = 0.22$ mA, estimate voltages V_A and V_C for V_S lower than V_A by 0.14V. [Hint: Assume that at the threshold voltage, half of I_{IL} is used internally.]

- **LS TTL Internal Characterization**

P2.1 Input Circuit for Input Low

(a) For the situation described in E2.1, measurements indicate $V_{A1} = 0.223$ V, $V_{A2} = 0.198$ V and $V_{A3} = 0.174$ V. Find an equivalent circuit for the input stage.

P2.3 Output Drive for Output High

(a) For the circuit of Fig. 14.2 with inputs low, what output voltage would you expect with a 10 kΩ load to ground? With 0.1 kΩ to ground?

P2.4 Output Drive for Output Low

(a) Measurement in E2.4 indicates that the output breaks from its clamped level of $V_C = 0.34$ V at $I_L = 21$ mA. For the circuit as in Fig. 14.2, with $V_{BE} = 0.75$ V, estimate the β of Q_3.

P2.5 Power-Supply Currents

(a) For all gate inputs high, the total supply current for an LS TTL Quad NAND is measured to be 2.6 mA. Estimate the total base drive available to one output device Q_3.

(b) Reevaluate β of Q_3 as supported by the data in P2.4

P2.5 Logic Gain, Edge Restoration, and the Schmitt Trigger

(a) For the circuit of Fig. 14.5, with a perfect input signal rising from 0V to 5V at node S, and the gates providing ideal outputs ranging from 0.4V to 3.6V, but with a 10ns pure delay, sketch the waveforms which result at nodes S, A, B, C.

(b) Reconsider the waveforms above for an input which rises from 0V to 5V in 11 μs. Assume the large-signal gain of each gate is –20V/V.

V EXPLORATIONS

- **LOW-POWER SCHOTTKY TTL – THE BASICS**

The internal circuit of a Low-Power Schottky (LS) TTL NAND gate is shown in Fig. 14.2.

Experiment #14–3

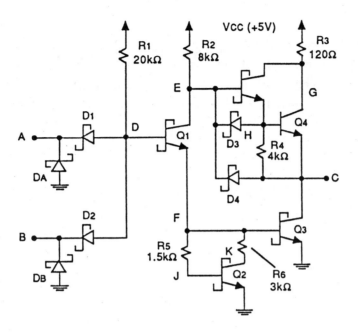

Figure 14.2 The Low-Power Schottky Circuit

E1.1 The NAND Logic Function

- **Goal:**

    To explore some basic properties of open-circuit inputs, and of the logic operation of the TTL NAND.

- **Setup:**

    O   Assemble the 74LS00 IC (Fig. 14.1) on your prototyping board, with power bypassing capacitors connected, (the low-inductance ceramic one wired quite close to the IC), and 5 V applied between pins 14 and 7, with pin 7 grounded. Leave all other pins disconnected.[1]

- **Measurement:**

    a)   Using your DVM with negative lead grounded, verify the supply to be 5.0 V, and measure the voltages at all pins, in triples (1, 2, 3; 4, 5, 6; 13, 12, 11; 10, 9, 8).

- **Tabulation:**

    Eight input voltages, output voltages.

- **Analysis:**

    Consider the pattern you observe: Distinguish the output and inputs. Note the (small) gate-to-gate variability within a single IC. (A second IC, if measured, would demonstrate IC-to-IC variability.) What is the significance of the voltage levels you have measured, in terms of logic-circuit operation? Attempt to justify the voltage values you find on the basis of the schematic provided in Fig. 14.2.

---

[1] Open inputs are acceptable with TTL, but not with CMOS. However, for improved noise immunity in a final design, unused inputs should be joined to V^+, possibly through a resistor for various reasons.

Experiment #14–4

- **Measurement:**

    b) With your DVM connected to pin 3 of the IC, ground, in turn, pin 1, pin 2, then both pins 1 and 2, noting the effect on the output.

    c) Repeat with other gates as desired, if you are not rushed fot time.

    d) Rather than simply leaving the inputs open, it is better practice to connect them explicitly to +5 V instead. If there is time, try this on one gate using 2 (somewhat long) wires, one for each input, connected either to ground or +5 V.

- **Tabulation:**

    Four output voltages, with two values each.

- **Analysis:**

    Consider what you have demonstrated: prepare a truth table in which H (high) and L (low) represents logic levels, and where an input is high when left open. Verify the logic function observed.

- **Measurement:**

    e) For practice in the application of logic gates, wire three of these NAND gates to create an OR circuit, and, then, four to create a NOR. Prepare a suitable sketch and Truth Table if you are not rushed for time.

- **Analysis:**

    Consider the relative economy of using a design process based on either NOR or NAND gates. Which is preferable?

E1.2 Operating Speed – Propagation Delay

When many inverting logic gates are connected in a ring, an oscillator results whose frequency depends on the details of the situation: If the number of gates is even, the frequency is zero, the circuit resting in one of two stable states (we shall explore this more later); If the number is odd, the circuit is more active, oscillating at a characteristic high frequency where each gate contributes its propagation delay for each logic transition to the period of the oscillation. Since each gate output must traverse from H to L and L to H in one cycle, the frequency of oscillation for n inverters (n odd) is f where $1/f = n(t_{PLH} + t_{PHL}) = 2nt_P$, for short. In the measurements which follow, you may consider using a frequency counter (if available).

- **Goal:**

    To use the ring oscillator structure to characterize gate propagation delay.

- **Setup:**

    ○ Connect a ring of three gates with the inputs of each NAND joined together and to the output of the preceding gate. Place one probe on the output (A) of one of the gates.

- **Measurement:**

    a) Measure peak voltages and the frequency at the output gate A.

    b) Connect the other probe to the output of the next gate (B), then to the next (C). Note the change in frequency with two probes connected, and the relative phase of all three outputs.

    c) With probes at A and B, measure the two propagation delays which characterize the intervening gate, using some convenient reference point, say half the signal swing.

d) Connect the fourth NAND gate, with inputs joined, to node B. With only one probe connected to D, the output of the fourth gate, measure the frequency, the peak signal value, and the rise and fall times.

e) Connect a second IC, if you have one, to form rings of five and, then, seven gates. Note the frequencies, and sample the relative phases in one or both oscillators.

f) With probes connected to adjacent gates, sketch and label the two output waveforms, specifically noting the propagation delays as measured at three reference levels: a) 50% of the signal swing, b) 50% of the power supply voltage (ie, 2.5 V), and c) the nominal switching threshold (say, 1.3 V).

g) Repeat at least part of step f) with probes connected to gates which are separated by a third (unmeasured) gate.

- **Tabulation:**

    f_a, f_b, t_{PLH}, t_{PHL}, f_d, v_{p+}, v_{p-}, t_r, t_f for rings of 3, 5, 7, and various definitions for t_P as appropriate.

- **Analysis:**

    Consider the ease with which average propagation delay can be found with a simple (ring) connection and measurement. Note the particular convenience possible with $n = 5$ (for which $t_P = 1/(10f)$), where, as well, signal voltages are quite pleasing in appearance, the output peaks being relatively near their static values.

    Consider application of the gate delays you have observed, possibly as delay elements in a logic network, or as a ring in creating a multiphase signal generator, for instance. As an example of the latter, first think about the output of a two-input NAND gate connected across three gates in a ring. Then, think about two such NANDs with the second set of inputs moved down the ring by one or two gates.

E1.3 Switching Threshold, Transfer Characteristics, and Input Characteristics

- **Goal:**

    To explore the use of a simple linear feedback circuit in measuring critical logic-circuit parameters.

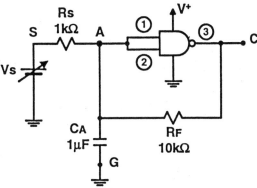

Figure 14.3 A Switching-Threshold Test Circuit

- **Setup:**

    O   Assemble the circuit shown in Fig. 14.3 with $V+ = 5.0$ V and $V_S = 1.0$ V initially. Note that the feedback connection shown makes it somewhat more convenient to adjust the input (and output) voltages using the variable supply V_S. Capacitor C_A, while often not necessary (since the feedback factor β is relatively low), ensures that the loop is stable by creating a very-dominant pole. (See Experiment #8).

- **Measurement:**
    a) With your DVM connected between nodes A and C, adjust V_S until $V_{AC} = 0$ V.
    b) Measure the voltages at nodes S, A, C. Estimate the gate threshold V_{th}, and the corresponding input current.
    c) While measuring from node C to ground, connect node A momentarily to 0 V, then to the 5 V supply.
    d) While measuring the voltage at node C, adjust V_S for a few interesting values of V_C, such as those very near, but not quite at V_{OL} and V_{OH}. Measure v_A.

- **Tabulation:**
    v_A, v_C, v_S under various conditions.

- **Analysis:**
    Consider the data you have acquired by preparing both a transfer characteristic (v_C versus v_A) and an input-current characteristic (the net current flow from the input node A versus v_A). Estimate values for V_{OL}, V_{IL}, V_{th}, V_{IH}, V_{OH}.

    Consider direct measurement of V_{IL}, V_{IH}. Recall that they are defined where the slope of the transfer characteristic is -1 V/V. How could you modify the circuit of Fig. 14.3 to obtain these results? Read on!

- **Goal:**
    To explore a circuit with which you can find V_{IL} and V_{IH}.

- **Setup:**
    O   Lift node G of capacitor C_A from ground, and connect it to the tap of a 1 kΩ – 10 Ω voltage divider driven by a 1 Vpp sinusoid at 100 kHz.

- **Measurement:**
    e) With your oscilloscope connected to nodes A and C, and set up to measure signals of 10 mVpp amplitude, vary the bias supply V_S until two values are found at which the magnitude of the small-signal voltages at nodes C and A are the same.
    f) Measure V_A and V_C for both, using your DVM.

    Note that the output may be quite "noisy" and erratic at the "gain of $-$ 1V/V" points. {More filter capacitors may help, such as a large ceramic one from S to ground.}

- **Tabulation:**
    v_a, v_c, v_A, v_C for special conditions.

- **Analysis:**
    Consider the fact that you now know quite a lot about the external properties of your LS TTL NAND gate.

- ## TTL Internal Characterization
    Our objective is to quantify aspects of the internal circuit of the LS TTL gate, including estimates of some of its component values, using simple external measurements.

Experiment #14–7

E2.1 Input Circuit for Input Low

- **Goal:**
  To observe properties of the resistive and junction components in the input circuit of the IC.

- **Setup:**
  - For one of the four NAND gates, connect the inputs together (as node A) to a 1 kΩ grounded resistor, leaving the output open.

- **Measurement:**
  a) Adjust the supply V^+ to exactly 5.00 V. With your DVM measure the voltage from node A to ground, V_{A1}.
  b) Lower the supply to 4.50 V, measuring V_{A2}.
  c) Lower the supply to 4.00 V, measuring V_{A3}.

- **Tabulation:**
  V^+, V_A for three values of V^+.

- **Analysis:**
  Consider what you now can find out about the internal connection to the power supply, and the values of the equivalent resistor R and junction voltage V_j. Use the fact that $(V^+ - V_j - V_A)/R = V_A/1$. To find V_j and R {where you could assume V_j to be constant as defined by a simple diode model}.

  Consider the use of the parameters you have just found, in conjunction with the results of E1.3 for operation at the input threshold, to find an equivalent circuit of the input stage of the gate.

E2.2 Output Drive for Output High

- **Goal:**
  To use some simple dc measurements at the gate output to characterize the important aspects of the output stage.

- **Setup:**
  - With the supply V^+ set to 5.0 V, ground one input of one of the NAND gates and prepare to measure its output voltage (at node C).

- **Measurement:**
  a) Find the output voltage V_C with the output a) open, then loaded to ground by b) 10 kΩ c) 1 kΩ, d) 100 Ω.
  b) With a 100 Ω load connected between node C and ground, reduce the supply voltage until V_C drops by 0.20 V, then measure the supply V^+.

- **Tabulation:**
  V^+, R_L, V_C, for various values of R_L for $V^+ = 5.00$, and V^+ for $R_L = 100$ Ω, and two values of V_C.

- **Analysis:**
  Consider the equivalent output circuit (as modelled by Fig. 14.2). Assuming that $R_L = 100$ Ω, forces the output stage to clamp or saturate to a fixed junction voltage, calculate a value for the resistor and junction component. Use the data for lighter loads. Assuming that $n=1$, for two

junctions what is the equivalent base resistance as seen through the emitter of the output transistor? Can you refine your estimate of the value of the current-limiting resistor in the collector of the output follower?

- **Step:**

  O Adjust the supply voltage V^+ to 5.0 V. Arrange a second supply, V_S, preloaded by a resistor of 100 Ω, and set to zero volts.[2] Connect a 100Ω load resistor from the supply V_S to the gate-output node C. Now, while measuring V_C, raise V_S, carefully noting that value of V_C for which $\partial V_C / \partial V_S$ reduces detectably. Measure V_S there as well. [Note that the change is not very great.]

- **Analysis:**

  Consider the significance of the measurements just made. What is threshold of current limiting? If, for the output transistors, $V_{BE} = 0.75$ V and $V_D = 0.5$ V, make a second estimate of the series limiting resistor.

E2.3 Output Drive for Output Low

With the supply voltage V^+ set to 5.0 V, connect both inputs of one gate to +5 V, and load the output with a 100 Ω resistor connected to a second supply V_S, initially set to 0 V. {No preload is needed.}

- Measuring the output voltage at node C, increase the voltage V_S until V_C begins to rise relatively abruptly. Measure V_S and calculate the corresponding current I_C at which the output transistor breaks out of clamping.

- Measure V_C at V_S equal to half the critical value just identified, then at $V_S = 1$ V and at $V_S = 0$ V.

Consider the output load characteristic by preparing a suitable plot of v_C versus i_C. Calculate the clamped output voltage and equivalent series resistance at a load of 10 mA.

E2.4 Power-Supply Currents

- **Goal:**

  To demonstrate the use of power-supply current measurements as a convenient tool for internal characterization.

- **Setup:**

  O Adjust your supply voltage to 5.00 V. Now, connect the Quad NAND IC to the supply via your DMM, connected to measure current (as a DCM).

Measurement:

a) With one input of all gates joined and grounded, note the current reading $I_{T\,min}$ quite carefully. One quarter of this is the average gate-input current.

b) Lower the supply voltage temporarily by 1.0 V, and note the new supply current, again quite carefully.

c) With the supply at +5.0 V once more, remove the common input connections from ground and carefully note the current. One quarter of this is the dominant part of the average current provided to the base of each grounded-emitter output transistor.

d) Lower the supply voltage temporarily by 1.0 V, and note the new supply current, again carefully.

---

[2] Note that the preload is intended to absorb current directed *into* the supply V_S.

- **Analysis:**

    Consider the significance of your measurements. Characterize the equivalent circuit of *each* input connection by identifying the corresponding resistor and junction drop. Likewise, characterize the basic drive circuit of the grounded-emitter output transistor. Use the combined results to estimate the resistor which supplies current to the upper follower part of the totem-pole output circuit.

    Consider the two resistor values just calculated, and using previous measurements (both in E2.3 and E2.4), estimate values of β for the two transistors in the output totem-pole.

- **Measurement:**

    e) With the supply voltage at 5.0 V, the DMM connected to measure supply current, and the inputs of three of the NANDS grounded, connect the input of the other NAND (pin 1) to its output (pin 3), *intermittently*, noting the *change* in supply current.

- **Analysis:**

    Consider this last measurement to provide an estimate of the peak current which can flow in the totem-pole transistors (from supply to ground) during switching, particularly with slow input transitions. What is the peak value per gate? per IC? Verify that these current values are reasonably consistent with estimates of β and resistor values deduced previously.

- **A TESTING TOOL AND ITS APPLICATION**

    Here we will explore the regenerative action of positive feedback, and two types of applications.

E3.1 Logic Gain, Edge Restoration, and the Schmitt Trigger

- **Goal:**

    To demonstrate a simple solution to the need for transition-time reduction: The Schmitt Trigger.

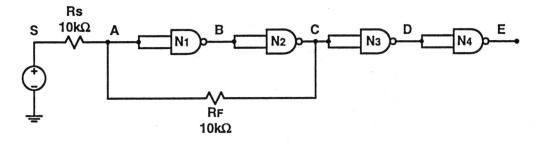

Figure 14.4 A TTL Logic Chain With Schmitt-Augmented Input

- **Setup:**

    O   Connect the four NAND gates of your IC as shown in Fig. 14.5, *but with R_F not connected*. Use a 5.0 V supply connected to pins 14 and 7 of the IC, with bypassing capacitors. As we shall see, resistor R_S serves several purposes, the immediate one of which is to isolate (and buffer) the waveform generator whose output is symmetric about ground. For ease of measurement, trigger your oscilloscope externally from the generator's triggering output.

    O   Connect a 3 V peak triangle wave at 100 Hz to node S.

- **Measurement:**
    a) Display the waveform at S on channel A, with those at A, B, C, D, E, displayed in turn on channel B. Note the transition times (10% to 90%) of each of the gate outputs.
    b) Connect resistor R_F and repeat. Sketch the related waveforms at nodes A, B, C.
    c) With R_F connected, lower the input amplitude until the voltage at node C no longer changes; then, raise it slowly to identify the peak voltage at S at which operation just begins once again. Note the transition times for the signals at nodes C, D, E for the input just above the threshold.

- **Tabulation:**
    V_{Speak}, R_f, t_{tA}, t_{tB}, t_{tC}, t_{tD}, t_{tE} for rising and falling inputs, for two values of R_F, and two values of V_{Speak}.

- **Analysis:**

    Consider the importance of logic gain G in providing an improvement in rise time as the signal propagates from N_1 to N_4. Convince yourself that for slow inputs the rate of rise of a gate output signal is $G \times$ that at its inputs. Using this idea calculate G_1 for N_1, G_2 for N_2 etc. But of course this process reaches a limit when the gate reaches its natural transition time limit where, conceptually, the input and output rates of rise of a symmetric gate are essentially the same. But in practice, few gates are symmetric, with equal rise and fall times. What about these? At what stage (N_1 to N_4) does improvement in transition time seem to stop? The reason that the Schmitt trigger is important is that its positive feedback (here, through R_F) allows a very large gain to be created with only two gates, allowing its output stage to operate at nearly the natural limit. Correspondingly, the signals at the output of a Schmitt Trigger are intended to be the fastest available in a logic family, essentially independent of the character of the input rate of change, provided the input is large enough. The application below will illustrate the use of this idea in the dynamic measurement of the peak drive currents available from a logic gate.

E3.2 Dynamic Output-Current Measurement

- **Goal:**
    To illustrate the benefits of a concern for transition time.

- **Setup:**
    O   To the circuit of Fig. 14.4, driven at 100 Hz by a suitable-sized triangular waveform, add a large load capacitor of 0.1 µF to node E,

- **Measurement:**

    a) Measure and sketch the resulting waveform. Estimate the greatest rates of change found in the rising and falling output voltages.

- **Analysis:**

    Consider this measurement as a relatively effortless means of estimating the maximum-available gate output currents, Note that the Schmitt Trigger ensures that the gate under test is driven in the most ideal way possible, and that, correspondingly, the peak currents measured are the greatest available.

Let us hope that a greater understanding of the inner workings of an important member of an important logic family will give you an edge over others who know only about logic design in the never-never land of idealized gates!

Experiment #14–12

NOTES

APPENDIX A
Experimentation

1	The Role of Laboratory Experimentation	**149**	*4* Testing	**158**
2	Laboratory Insights	**151**	*5* Troubleshooting	**159**
3	Experiment Layout	**155**	*6* Safety	**161**

A1 THE ROLE OF LABORATORY EXPERIMENTATION

There are two opposing views of the potential role of laboratory experimental work in the process of education. The first, and more conventional, is that laboratory work is a means to confirm (and thereby reinforce) knowledge gained in other ways (such as in lectures) – **the reinforcement view**. The second, and less understood nowadays, is that laboratory work is a means by which one samples the nature of a physical reality, in order to provide a basis for insight to be gained later by some complimentary activity (such as by analysis, or subsequent experiment) – **the experience view**.

Regrettably, the programs to which many students are exposed today, commonly emphasize only the first view. Indeed, that view, as one of a process of **reinforcement**, seems entirely logical, thus entirely laudable. However this view is entirely too simplistic, even simple-minded, for it reinforces, as well, the false notion that knowledge is available only *in neat packages* – from lectures, from books, or from some other source of **authority**. Unfortunately, this emphasis on formalism seems also to serve to denigrate the worth of the second view, that of the importance of simple **experience**.

This situation is in fact a manifestation of a quite general problem in today's society – the somewhat overwhelming sense that everyday technology is so complex as to be beyond the understanding of all but the **specialist**. Thus to learn about engineering, one goes to an engineering school for a "formal" education provided by **authorities** in various topics, **experts** in their fields. The idea of "experts" is continuously reinforced around us – on TV by talk-show hosts, in the home by plumbers to whom we are willing to pay astronomical sums for presumed expertise.

Today, the perceived need for expert knowledge is so pervasive that it is increasingly rare for an individual to even attempt to repair usual times around the home. Yet at the same time, external experts are expensive. Thus the norm is to scrap most movable objects in favour of new ones. In fact, in the case of the plumber, one might cynically note that one pays the expert's ransom, primarily, and simply, because the plumbing is not normally portable!

Thus in relatively affluent societies, the conjoined notions of repair, and of being handy with tools around the home, have virtually disappeared. And with this change, the term **tinkerer** has evolved once more, to become almost an insult.

Other facts reinforce this idea: For example the pervasive term *"trial and error"*, used to describe what may in fact be an organized iterative process, carries with it a negative connotation. As well, by an unfortunate pairing of terms, it reinforces the idea of expert knowledge as somehow more **legitimate**. Now, in the latter context, consider the potential role of a more positive view, one of which we can all be proud – the process of *"trial and success"*.

It is to this latter procedure, **trial and success**, that your present laboratory experience should be directed, for in 3 words it captures the potential role of the laboratory in the process of initial learning (rather than only of learning reinforcement).

While the laboratory also serves as a place in which to practice what you know, such a view, alone, is very dull. Moreover, and worse, it is very limiting, for the very concept "know" implies "know well" or perhaps (only) "know well-enough". Thus with it comes a reluctance to explore, to deal with the little-known. Regrettably it is being sucked into the trap of the need for *certain knowledge* that leads the simple-minded to cheat (themselves), and rely on other's work, when driven to substitute grades for understanding.

So, rather than simple reinforcement, our goal here is **exploration** at, or beyond, the limits of our certain knowledge. In this process we wish to make use of *any* relevant experience to gain more! Our tools will be simple general-purpose ones, such as analogy or equivalence, the identification and modelling of generic properties, and the formulation of general-purpose somewhat-universal questions. Such questions themselves are examples of the experimental process of the second kind, that is of exploration into the unknown from a tenuous basis of incomplete knowledge.

In Summary: The laboratory experience is all about *learning* rather than about *teaching*. In particular, all that can actually be taught there is about the generic process of **learning to learn!**

Correspondingly, it is essential to learn that the **laboratory-learning process** is an active and participatory one. It operates in a closed-loop cycle in which results are understood and assimilated "**on the fly**", so to speak. Mere data collection for subsequent analysis by experts is better done by technicians, and perhaps best done, nowadays, by automated instruments! The essence of the process of laboratory-learning is highlighted by the description of its major activity as "**hypothesis and test**". It is this phrase which captures the cyclic action of **confirming a conjecture** as a means of positive reinforcement!

Remember:

Learning to Learn!

Hypothesis and Test!

Confirming a Conjecture!

Trial and Success!

A2 LABORATORY INSIGHTS

• The Laboratory Environment

It is useful to view the laboratory (whether academic or industrial) as a learning environment, in which activity can be **structured** for vastly increased efficiency. To this end, the challenge you face in performing effective work in the laboratory can be captured as a particular procedural paradigm, namely, **Prepare; Process; Preserve; Ponder; Present**. Propping up these preferred procedures are profuse principles, partially profferred below:

- **Preparation** – Organize your task ahead of time to reduce non-productive thinking time, and to avoid false moves, during the heat of laboratory action.

- **Patterning** – Establish a general pattern for your activity. Don't keep reinventing the wheel! Establish an efficient process, and use it until you find a better one; but don't miss an opportunity to improve!

- **Progression** – Generally speaking, proceed in an orderly fashion: Act for a purpose; savour stability; cherish continuity. Normally, change only *one thing at a time*, particularly when trouble-shooting. Trouble is all about unusual behaviour. One possibility is that *you* are the cause – *you* don't understand what *you* are doing. Work in sequence – single conclusion, single change, single verification.

- **Indolence** – Be creatively lazy! Minimize your effort through *cunning*. Don't take unnecessary steps. Think before you act (that's what *preparation* is all about). Recall that the laboratory experience is a cyclic one, of **hypothesis and test**! Hypothesize before taking action!

- **Awareness** – Maintain your focus on the job to be done. *Think* as you proceed! There is a lot of information passing by in the laboratory process. Try to be aware of it, and try to use it. Thereby, avoid mistakes; learn more; discover!

- **Checking** – Develop the habit of checking as you go. You know, or will shortly know the importance of feedback: Close the mental loop around your physical activity!

- **Stability** – Proceed one step at a time with respect to some *constant* unchanging reference. Be *quick*, but *not hasty*!

• Getting Ready – The First Step to Laboratory Success

The Nature of Preparation

Your laboratory experience can be vastly rewarding, or abysmally depressing and apparently futile, all depending on your expectations, and what **you** do in their support! We all hear of some students (and people in general) who are said to be "lucky". Things seem to come easy to them! But is this a fact, or a mere delusion?

I personally favour the latter view, for I believe strongly in a comment I once overheard on Japanese TV (in English!!), to the effect that *"luck is the encounter of preparation with opportunity"*. Reflect on that! Bear it in mind when the pressures on your time make time for preparation low in your priorities. Be aware that opportunity is always around: "Opportunity knocks", so they say. But who is listening? Who hears? Probably the *lucky*, those who are prepared!

As you participate in the laboratory experience, bear all this in mind. A laboratory is potentially a special place. In the sometimes forgotten traditions of science, it is a place for **discovery**. But discovery is an *active* process, a process best performed by a mind prepared in expectation of the new and exciting. The laboratory is not a place to prepare, but rather where one benefits from *prior* **preparation**. If there is no time;

Appendix A–4

so be it. But be aware of this choice, and its potential for unfulfilled expectations!

Being Organized

Be prepared! In addition to the obvious reading, thinking, or analysis that should precede any useful process, such as a session in an academic laboratory can potentially be, there are some things which, done once and for all and early, make your life less of a hassle!

- In general, don't carry stuff in your head when it can be better written down, thereby facilitating checking and rapid test.
- Gather data for your components. Questions will inevitably arise for which a good answer, quickly, saves time.
- Prepare and organize the actual components to be used: This includes wires for connection. In a perfect world this would include testing of components ahead of time, as well. If that is not possible, having spares of each important element is a good idea. Obviously measuring resistor values is trivially done, and a good idea if there is any doubt about colour coding, etc.
- Prepare **explicit** circuit schematics for each circuit to be tested, with *labelling* of device pin numbers, etc.
- Prepare the circuit schematic to fit the prototyping-board assembly process. Use the power-rail-related arrow notation for the power supplies, and orderly signal flow from left to right. Organize the circuit drawing so it is "flat" in the sense that a prototyping-board layout must (nearly) be.
- Consider providing a (rough-and-ready) layout suited to the prototyping-board of (big) circuits in which, for example, best choices of IC internal elements are chosen by physical pin location for improved layout ease.
- Where feasible, use packages with single devices rather than multiples, for ease of layout.
- Wire around IC's in such a way as to allow them to be conveniently replaced, as required by the **Experiment** or as part of trouble-shooting.

• On the Bench – Techniques for Effective Measurement

Wiring a Circuit

Work from a complete schematic: Use a drawing convention with conforms to normal wiring practice and topology (with signal flow from left to right; with bused power rails; with positive up, etc).

- Consider a layout sketch.
- Check off connections as you make them (or trace them using coloured pencils).
- Count generic components.
- Count connections to each node.
- Bring out special nodes to test points (using a resistor link).

Connection of Generators

In general, be aware that signal-generator outputs are often unsatisfactory for direct use in circuit testing. Occasionally the output voltages available are too small, or too little current is available, in which cases your experiment needs an additional amplifier. More likely, however, is that the lowest available output is too large, and overloads the input (for example, of a high-gain amplifier). Thus, generally speaking, include a simple resistive input attenuator (using, for example, 1kΩ resistor in series from the generator with a 100Ω (or 10Ω) resistor shunted to ground) at the input of your experimental circuit.

Measurement Order

Make measurements in a logical order. (Recall the basic "idea": first hypothesize; then test.)

- Relate each measurement to the past: Build up a complete picture. No one measurement holds "the answer".
- Be orderly: For example, as a quick check, measure input, then output, looking for constants of the process (eg, waveform, frequency, etc).

Shunt Measurement

In general, recognize the obvious fact that it is far far easier to connect to a single node of a wired circuit and measure its voltage than it is to open a branch and measure the current there. The degree of disruption to the original circuit is enormously different in the two cases. Thus, in general,

- Use voltage measurement (shunt), *not* current measurement (series).
- Make voltage measurements single-endedly to a common reference.
- Use differential measurements only when difference matters in great detail. Otherwise, calculate (mentally) the difference between the results of two separate measurements.
- Ensure that the available measurement lead is appropriate − don't use the probe or wire provided if it is too unwieldly. Rather, use a mechanical impedance-matching scheme, an intermediate anchor point for attaching a large wire to a small one. For example, a resistor (of a few kilohms), connected to the instrument with an extended flexible lead on the instrument end, often makes a good probe. One can hold the resistor body, as an insulated handle. One can have the probing lead (the resistor's other end) as long or as short as desired, and insulated (with "spaghetti") or not. As well, the series resistance isolates high-frequency aspects of the circuit from long meter leads.

Oscilloscope Measurement

- Generally speaking, tend to use **oscilloscope probes**, either fixed ×10 ones, or switchable ones set to ×10, for most measurements. Use ×1 probes sparingly, and only if the additional gain is *absolutely necessary*, since such probes present a very large capacitive load to the circuit under test, and other problems.
- If available, use 2 *(or more) oscilloscope channels* with "alternate switching". In best practice, one of these channels displays the signal held at a fixed location in the circuit, while the other displays the input to a roving probe. For initial tests, the fixed probe is best placed at the input, with the other is first used to measure the output (to see if the circuit works at all), and then to proceed from the input to successive nodes to see why there is a problem. (If you were absorbing what you read just now, you will have noted that having both probes on the same node initially, allows one to check the oscilloscope channels for consistency.)

  Alternatively, for a circuit that already works, whose detailed operation you wish to examine, placing the fixed probe on the output is often a good idea. This is the case since it allows you to detect whether probing internally in the circuit affects it critically, in such a way as to modify its operation. With only a single channel, there is a real danger that (as Heisenberg has suggested) measurement will affect operation. A classical case is that of a bistable element having nominally complementary outputs. Such a circuit, if sensitive to probing, often appears to have failed, that is to have identical outputs when viewed by a single channel instrument, one output at a time!

- Generally speaking, try to use external triggering. Triggering from the "fixed" input channel is better than nothing, but second best, since usual changes made in the process of experimentation, for example, varying input amplitude, affect triggering operation. Ideally, triggering should be directly from a separate constant-level output from the signal generator being used, or, lacking that, an equivalent arrangement created on the prototyping board, with a potentiometer installed in the input-signal path, where variation is needed.

Oscilloscope Display

- Ensure that the displayed time scale is appropriate: Typically it should include more than one cycle of a repetative event, but not too many.
- Usually, maintain the dc component in the display unless, or until, the changing part is most critical, and can be emphasized by the ac-coupling option on your oscilloscope. Certainly, to begin, use dc-coupling of scope channels in order to establish the basis for an ongoing complete picture, one whose aberrances might trigger you early to a concern for potential trouble.
- Generally speaking, operate the vertical channels at fixed gain (that is, with the variable knob at its calibrated end.
- Usually, separate the channel traces on the screen to facilitate unambiguous interpretation; *but* ensure that the reference level for each channel is established at a convenient place, and kept in mind.
- Occasionally, both (or all) traces should be shifted to have the same reference level to facilitate detection of small changes. In such cases, both channels of the oscilloscope are often operated at the same gain setting to facilitate direct comparison. To check matching of channels, connect each to the same circuit node. Adjust the vertical separation controls and the gain control of higher-gain channel, to ensure complete identity of presentation. (Be concerned if the adjustment of gain control is too great, a sign that oscilloscope repair is needed.) This is the process called "**normalization**" described more fully in Appendix B3.

Remember something which guarantees success in laboratory work:

Luck is the Encounter

of Preparation

with Opportunity!

A3 EXPERIMENT LAYOUT

As noted earlier, it is assumed that the circuits you will explore in the **Experiments** provided in this **Manual**, will be assembled on prototyping boards as described in Appendix A4 *Generic Equipment*. For convenience in this segment, (and later) we will use the mnemonic label PB or simply the term *board* rather than the more complete but unwieldly word pair.

It is important, for your effective use of you own time, that you realize that the pace of your experimental work will be affected greatly by the style in which you assemble circuit components on your PB. In general, planning and structuring are important. An unplanned board layout, created randomly and incrementally, without some unifying scheme, can become very chaotic as the circuit ultimately enlarges. Such chaos leads in turn to mistakes in interconnecting, possibly damaging short circuits and general reduction in the effective pace of your work. Though you are very busy coping, you are likely going nowhere!

Again, very generally speaking, the alternative idea is to make your assembly follow standard patterns as much as possible. Find a pattern that "works for you" and stick to it. Employ schemes that are inherently mnemonic and embody, and enable, a process of inherent self-checking. As we will see, the recommended arrangement of power supply connections on properly-located power rails or buses is such a scheme.

More specifically:

- **Begin with a well-labelled circuit diagram**: Yes, for simple circuits you may not need it, but there are better things to remember in the heat of battle than circuit interconnections. In general, you will see the situation clearer on paper. As well, you can exchange ideas with your laboratory partner or Instructor more effectively: It is simply much easier to say "this (resistor) while pointing, than to say "the 10kΩ resistor connected between the collector of the first transistor and the base of the second".

- **Establish and maintain a global pattern in the layout of your PB**:

  - Signal flow should be left-to-right generally, if possible.

  - Physically modularize your circuit into functional blocks in which you clump associated components together to the degree allowed by the types of components you have, and the rules your Instructor provides concerning cutting of component leads, and so on (See *Treatment of Component Leads* in A4 above).

  - Locate power-supply rails in an orderly fashion: For a ± supply system try to arrange positive supplies at the top, negative at the bottom and ground in the middle. For multiple supplies of one polarity, try to arrange their buses in mnemonic order, the most extreme voltage to the outside of the board. As well, in general, try to arrange as large a ground-bus network as possible. Certainly, if it is not possible to have a ground in the middle, have one at each side the board, preferably innermost. In spite of what you seem to hear about "ground loops" and the like, much of which is quite spurious and forgotten-context-laden, join your multiple ground buses together with direct (black-) insulated-wire connections.

- **Consider a layout or wiring diagram, distinct from the circuit schematic:**. All are obviously related, the schematic being of necessity the most complete. The layout diagram simply locates the most significant and critical parts and labels pins. While logically a planning aid constructed prior to wiring, and used for that process, it also serves to document what was really done and allows a later observer (yourself or your Instructor) to interpret certain kinds of results related to noise, interference, spurious oscillation and the like. The wiring diagram, at its most complete, includes a representation of the layout with all components roughly to scale with lines respresenting wire interconnections. To produce a wiring diagram is often a lot of work, appropriate only in critical situations, possibly where lead lengths are critical, signal proximity is an issue, printed-circuit fabrication is contemplated, and so on. Often a layout diagram with a few critical wire routes noted, is useful and sufficiently quick. Certainly, until you develop your own dynamic PB-assembly style, a partial wiring diagram is a good idea. Planning usually is!

Appendix A–8

- **Record connections as you make them:** Lacking layout and wiring diagrams, or even with them, it is often a good idea to note connections installed, as you make them, on your schematic, using simple marks or line tracing with coloured pencils. To resolve the one-to-many mapping between the schematic and the wired result, which a wiring diagram is intended to resolve, use dominant components (ICs, coils, transistors, possibly capacitors, but *not* resistors) as the reference points. Follow the priority order implied in the list above, viewing the pins of the major components as reference points. Try to associate minor components (resistors) with a major one by related function. Thus a base-bias resistor for Q_2 is near it rather than near the collector of Q_1 which drives it. Occasionally it pays to create a local "node" to which minor components are connected, which is then linked to the major component (say an IC). The end of a capacitor is often an appropriate location for this node. More is said about this in conjunction with multipin-package (IC) wiring styles following An important related issue is in what order to connect the minor components, and when to connect the interconnecting wires. Obviously if the wire is an essential *permanent* part of the topology, it might logically be connected first, perhaps low-down on your PB. One reason for this is if you *ever* want to remove a wire, the (minor) parts connected originally to each of its ends are evidently treated quite differently. One approach to this is to substitute small resistors for all wires, in which case the decision as to end, is forced on you at an early stage. Note that if you use this technique, the schematic should be notated with the location of the resistor connection. While a more explicit notation may occasionally be necessary, often a single arrow to a line with a label indicating resistor value (eg, 24 for a 24Ω resistor with red, yellow and black bands) is more than enough. Of course, this issue is what a *wiring diagram* is all about!

- **Be aware of distinct layout styles:** In general, the style of layout on your PB, and the use of schematic layout and wiring diagrams, and even of wiring lists, depends on detailed circumstances:

  - If the setup is dominated by IC packages, as experiments in digital logic normally are, a layout diagram may be useful, where the physical layout and the layout diagram correspond geographically. Depending on the context in which it was created, that logic-block schematic may be usable directly. In general, layout position choices are made on the basis of global functionality, interconnection patterns, numbers of connections, and wire lengths.

  - If the setup has a large number of relatively small packages, each with a small number of pins, particularly discrete components, capacitors, diodes, transistors and the like, a schematic circuit diagram is likely to be most appropriate. Here the physical layout follows the schematic topology, with the anomalies related to the occasional use of a few multipin ICs accommodated by the use of local jumpers to resolve the topological mismatch between the schematic and pin connections. With this arrangement, the circuit-drawing style employed throughout the **Text** and this **Manual** correspond well, with the recommendation of power-rail location on your PB, namely positive rail at the top, and so on. In both cases it is usually best to orient all multipin ICs the same way in order that corresponding pin numbers appear in coresponding places. This tends best to fullfill the need for rapid pin location, with characterizes most experimentation styles. However there are obvious exceptions: They generally occur as special cases when there are very few (particularly only 1) of some particular device. This occurs in digital applications with some very large ICs (for example), and in small linear systems using op amps. Generally speaking, if there are two or more of the same IC, their orientation should be kept the same. Otherwise errors in initial wiring and/or measurement characterisically result.

- **Organize Mechanical Aspects of the Test Setup**: As is done on some complete commercial prototyping systems, it is generally a good idea to arrange that there are physical terminations for connections of external instruments to your PB. Thus a metal panel on which your PB sits can include some useful auxilliary facilities.
  - Terminals for the connection of power supplies.
  - A multipole switch for isolating all supplies from your board for major (if not *all*) component changes.
  - Ground terminals for the connection of oscilloscope ground clips.
  - Terminals for the connection of DVM leads. In all of these cases, the terminals, connectors and so on provide a mechanical "impedence match" between unruly instrument leads which tend to physically overpower and overturn an isolated PB. They are usually connected to some standard pins on the PB from which connections are made in the usual way. Note as well, that a "standard" facility on your PB can be large electrolytic or tantalum capacitors for power-supply bypassing, with capacitors for high-frequency bypass placed close to the actual circuit, and possibly sectioned (digital versus analog), by small series resistors (or chokes) in the power-line connections.

- **Use Colour-Coded Wiring**: To make troubleshooting easier, and to reduce the chance of misconnection, and, further, to make such errors more detectable, try to use consistent wire colours: Black (or possibly green) for ground; Red for the positive supply; Blue for the negative supply; Orange and purple for lesser positive and negative supplies, respectively; and yellow and white for signal connections.

A4 TESTING

As the complexity of electronic circuits increases, the process of testing becomes ever-worse. At present, testing often constitutes the overwhelming component of a total systems-development cost. Accordingly, "design for test" is a very important aspect of modern electrical-engineering activity. As noted elsewhere, modularity in design and/or in testing is a good idea.

Meanwhile, you do whatever else you can to make the process of test as effective as you can manage, at lowest-possible cost. To this end, there are a few **basic principles**, which, if understood, make testing a lot easier. Some of these, which are particularly relevant in a learning-laboratory setting, follow:

- DC measurements are easier than signal measurements. A corollary is that they are often less informative as well; but so it goes!

- Voltage measurements are easier than current ones, because they imply less impact on the circuit, no leads being broken, and so on. (Obviously, Heisenberg intuitively knew this fact.)

- Single-ended-measurements (between a single node and a common point) are more than-twice-as-easy-to-make as differential measurements (between two nodes). Thus, keep one end of your DVM or oscilloscope connected to a common point, typically (or nominally) ground. Be prepared to do some rapid mental calculation to obtain difference data. Note that modern digital oscilloscopes with memory make the process of direct subtraction of two separate single-ended waveform measurements quite easy.

In summary, and to generalize, more is to be learned from a test when it is done quickly! In the limit, if you take too long, something else will go wrong, or at least change! Be quick! In particular, trouble-shooting done slowly, is a multiple loser: Too much time is spent on something that shouldn't have been needed, at all; and, more importantly, while you are doing something that should never have been needed, *you* are *not* doing something useful. A great waste, in a busy life!

A5 TROUBLESHOOTING

Often, circuits when newly wired, fail to operate as expected. There are many possible reasons; your challenge is to find *all* of them efficiently. Notice "all" and "them"; the assumption of a single fault is often a major stumbling block in troubleshooting, a classical failure mode of the troubleshooting process itself. *Expect multiple faults*; if you were wrong or unfortunate once, why not twice?

Some of your problems originate in the prototyping-board environment with which you have been provided: contacts lose their gripping ability, wires/components are not correctly inserted; you have an incorrect view of "hidden" connections. As well, there is generally a lot of potential for "miswiring". Obviously, spend some time checking your wiring (ideally as you go), *but* don't waste too much time looking for the illusive "*something*": Rather, take an "**activist view**".

One *wrong way* to proceed upon detecting trouble is to rip the circuit all apart and start again. While this approach is all-too-common, it is wrong-headed, regrettably representing an adherence to the view of electronics as "magic", one which is certainly out of place in the present context.

A proper "**activist view**" is one of careful, ordered, but rapid, checking of the situation *in a logical way*, using a single external probe attached to an appropriate instrument (but possibly using a second probe and instrument as a fixed "reference"). The task ahead is one of *verifying your expectations*, on the way to locating where something is wrong. For troubleshooting, you may often use a DVM, because of its versatility, economy and portability; however it is often useful (and sometimes necessary) to use an oscilloscope. That the scope comes with an almost-insulated ×10 probe which offers inherent isolation (decoupling) from the circuit, is often an advantage. Certainly, there are some kinds of problems that can be found *only* by using an oscilloscope.

In any case, the first thing one checks is the instrument itself, and its connections: Instrument common should be connected to circuit common, and only a single instrument lead should be moved *at one time*.

- First, having adjusted the instrument range appropriately to measure dc in the range expected, measure the non-common end of one power supply (often called a "powerrail" or, simply, "rail"), preferably near its source. By this single act, you simultaneously verify the likelyhood that the instrument works, and that its calibration is reasonable, as well as that the power supply works, and that its voltage is also reasonable.

- Proceed then, past switches, to *all* (or a sample of *all*) of the individual places that power is supposed to flow. Repeat for other power rails. Note that probing on the ends of *all* resistors connected to a rail can verify that they are connected, and at the same time you can verify the count of all the connections (using your schematic) and, most importantly, that components are connected as expected, to the rail. Note however, that such *exhaustive checking* can be quite *premature*. It is certainly appropriate when desperation strikes, but you are not *yet* desperate. Otherwise, it is too procedural, too robot-like, too intellectually passive. A more **active** approach is better, particularly at the early stages of the troubleshooting process!

Now that power's presence is proven, two somewhat-separable procedures are possible. They both involve **tracing**, a process in which the general idea is to proceed from one node having a known signal to the next, to verify the existence of a predicted value, in the **active** process of **hypothesis and test**. The first uses dc as a "signal"; the second, more conventional, uses ac.

- In the dc-tracing process, the power supply is viewed as a signal. Generally speaking, the idea is to sample circuit nodes, verifying that their dc voltages are reasonable. There are many possibilities: one approach is to look first at the pins of ICs in the circuit (or of any other major elements). Another is follow along component strings to check all nodes in connected order; however in its extreme form, this is, generally-speaking, also *too procedural*.

Appendix A–12

Another, somewhat-more-sophisticated approach, is to look at the dc voltages on each side of every discrete capacitor visible. And so on, and on ....

While it is quite remarkable what can be learned by some of the "dc" processes above, they eventually encounter difficulty when the circuit is driven by signals, or, for its own internal reasons, is active itself. This is not always, the case, but often, particularly for nonlinear operation. Thus, shortly, an oscilloscope becomes necessary, but to use it in a direct-coupled (dc) mode is often still best at this stage. There are, however, valid arguments which say that a signal-only (ac) trace is quickest at an early stage in troubleshooting.

- For input-signal-based tracing, typically start at the input generator, proceeding along the connected signal path to verify relevant signal properties (typically, only general wave-shape and amplitude at first). Note that to make this process useful you must be able, very quickly, to estimate *appropriate* signal transformations between nodes. The process, in general, involves rapid analysis, or alternatively, off-line, precalculation.[1] Sometimes, the input signal is too small to measure with the equipment you have. However, often, it is possible to raise the input level temporarily, to "see what happens". Occasionally, this is not possible, or may even be dangerous. In this case, your only recourse is to jump further along the signal chain to seek a signal (or to detect its absence), and then to work *backward* to where the problem originates.

When the usable input is too small to measure, yet you cannot find a signal farther along the signal chain, one possible technique to explore is to measure at the output again, while inserting (large) signals from an external generator, via a series capacitor and resistor. You include the capacitor so that the bias is not disturbed, and the resistor to limit signal currents, thereby preventing possible damage. **Never** connect external voltages or signals directly inside a (typically poorly-known) circuit.

One additional technique worth mentioning for use during troubleshooting, is "node shunting". It can be used to quickly characterize the impedance of a circuit node at which the signal you find is somehow too small. The idea is simply to shunt the node in question with a load, typically the series combination of a large capacitor and a resistor, while observing the voltage at it or at some node to which it provides the signal. The capacitor, typically large and grounded, is used to isolate dc. The resistor is reduced until the value which lowers the signal by half is found, providing a direct measure of node resistance.

While these comments are certainly not exhaustive, they may help to establish the active up-beat mind set which is essential to a rapid and effective counter-attack on unruly reality.

---

[1] Again this is often too procedural, too cold-blooded, too failure-acceptance-oriented, to provide the exhilarative "up" mentality on which successful troubleshooting can flourish.

Appendix A–13

A6 SAFETY

This issue of safety in the Laboratory has two components, one **personal**, and one **physical**. Fortunately, the personal environment of an undergraduate laboratory in electronics is relatively benign: The instruments used are dominantly commercial ones whose power-line related integrity, while not assured, has been examined by standards organizations like UL and CSA; The power-supply and signal voltages recommended in this **Manual** are less than 25V peak amplitude, and the power levels are at most a few watts. However there are more subtle issues of which to be aware:

- Check the instrument line cord. The wire and plug should be in good physical shape, not kinked or broken, and the plug should have a functional ground pin attached to the instrument case, (unless that is well-insulated with no metal parts available to be touched).

- Be wary of power-line operated apparatus of a non-commercial variety, for example the normally low-voltage transformer used for possible ac experiments. Before using such apparatus, check if accessible metal parts are grounded. If some part is not obviously so, use your DVM connected between it and local ground (for example the frame of a plugged-in commercial instrument (*but* check with the ohmmeter *between* two such instruments to be sure they are interconnected through ground), to detect any possible ac voltage. With your voltmeter connected, reverse the line plug of the non-commercial apparatus, if that is possible. If you find a voltage, shunt the DVM leads with resistors to discover the source impedance and possible current level. **Be careful**; Consult your Instructor.

- During the **Experiment** itself, be wary of overheated components which might burn your fingers. Before grasping an unknown component, test it tentatively with one (expendible) finger tip. Until you are sure that the power level is quite suitably low, be very careful in grasping a component while connecting it to an active circuit. If in doubt, at least first time, remove the power first, then connect; then repower.

- Otherwise, be wary of sharp wires, sharp pins, (especially when inserting DIPs in your board), and heavy instruments. *You* are not indestructable! Your partner is typically even more delicate!

Concerning the protection of physical property, the instruments and components you use in these **Experiments**, there are only a few issues of which to be aware:

- Generally speaking, use the current-limiting capability of your power supplies to the *fullest possible extent*. Ideally the limiting level is adjustable, in which case *keep it low*, adjusting it upward only as the application requires (as signalled by the appearance of an inadequate voltage level).

- Corresponding to the concern expressed above for finger burning, be aware of the power ratings of components. Generally speaking if a component burns your finger, it is up to no good! Know the reasons! If you do not, turn off the power, and reflect!

- Be aware that charge stored on a capacitor can be dangerous to other components. Thus when you pick up a capacitor, particularly a large electrolytic one, short its leads to discharge it, safely away from other components. Do so twice, (or even more) for several reasons, including the fact that leads may be oxidized, that redundancy is good in general, and that (more subtley) a (re)polarization charge-redistribution phenomenon in some capacitors allows some charge to be retained upon each discharge.

- Be aware of the effect of electrostatic discharge on semiconductor components. MOS devices are particularly sensitive, though junction devices are not totally immune. One problem is that internal damage may not be immediately evident. If there is overt evidence of static-electricity generation in your environment, then the situation is *very bad*. Consult your Instructor. Change you clothes. Get a new lab partner. Stop shuffling on the carpet! If all else fails, and even as a good general practice when coming up to your experimental setup, first touch the case of a large instrument (the oscilloscope say); then sit down, then settle down, then touch the instrument again; then proceed, but *don't shuffle*.

Appendix A–14

- It is good general practice to keep ICs, particularly MOS, on a slab of conductive foam. Be careful, since regular foam can be dangerous to some components, however handy.
- Be careful while inserting DIP components into your prototyping board, not to bend or compress the pins.
- Be aware that electrolytic capacitors can explode if connected with incorrect polarity in a circuit in which current is *not limited*. (This is one of the many reasons for using a current-limited supply.) Note well that the outside metal can of an electrolytic must be kept more negative than the other (isolated) terminal, at least on the average.
- As well, there are other precautions relating to component safety, involving excess signal-generator voltages, injection of voltages etc, of which more is said in association with a description of good experimental technique appearing in Appendix B1 and B10.

BE CAREFUL OUT THERE!

APPENDIX B
Instrumentation

1	Power Supplies and Current Limiting		163
2	Oscilloscope Calibration		165
3	Oscilloscope Measurement		166
4	Differential Measurement		168
5	AC Measurement		170
7	Linearity Distortion		172
8	Temperature Testing		173
9	Roles for Potentiometers		174
10	Roles for Resistors		176
11	Control of Parasitic Oscillation		177
12	Standard Component Values		180

B1 POWER SUPPLIES AND CURRENT LIMITING

In this section, we will consider various aspects of appropriate regulated laboratory power supplies, their packaging and their controls, particularly the ongoing use of their controllable current-limit facilities. The most appropriate and versatile power supply for experimentation in electronics has an isolated variable output voltage regulated against both line and load variation. The output voltage should be adjustable over the range from *zero* to some characteristic maximum voltage (say 25V for the present purposes) and essentially independent of the load current, up to some current limit. This current limit is ideally also adjustable, from essentially zero to a value representing the maximum curent output available (say 500mA for the present purposes). While all of this is ideal, there is a marked tendency, for economic reasons, to package supplies in multiples, with a common ground connection. One such commercial triple supply unit, for example, provides ±18V and +6V, all with variable voltage, the latter as a logic supply. Another trend is to fix the current limits at the maximum rating of the supply. Such is the case with the +6V supply in the triple unit referred to above. Unfortunately, though a fixed current limit protects the supply, it does nothing to protect the load. Thus a supply with a fixed current limit does only half the required job. Note, by way of example of the problems with multiple fixed formats, that the triple one described cannot deal with mixed analog-digital systems using ECL logic for which a −5.2V supply is necessary.

To cope with the possibility of this type of problem, the **Experiments** in this **Manual** use a ± supply with common ground exclusively, with a third supply of either polarity being a convenient replacement for the specified alternative potentiometer connections. For your convenience, Figs. E9.1 and E9.2 in Appendix E9 provide relatively simple circuits which can be used to create a third low-current regulated supply, whose voltage can be adjusted over the range of ±7.5 V, from the basic ± supply pair. The circuit in Fig. E9.2 is also equipped with a fixed current limit.

With supplies having both adjustable voltage and current capabilities, a relatively orderly procedure is available for testing a newly-wired circuit:

- With the circuit disconnected from the supply, turn the supply on, and adjust its voltage to the desired value, as measured by an external DVM. (You can check the supply's voltmeter at this point as well, if such is provided.)

- Now, lower the current limit to check characteristic behaviour: For some designs, at the lower end of the current-control range, the voltage will drop, even with no load. Leave the current control as low as possible while maintaining the desired output voltage.

- Short the supply through a small test resistor, say 10Ω for our purposes (of which more will be said later), across which your DVM is connected, to check the maximum available current.

- (Be aware of the power rating of your test resistor.)
- Now adjust the supply to provide the estimated maximum current required by the circuit.
- Having repeated all of this for each supply, you are now ready to switch on your circuit.

Note that in the previous description a small resistor is used. It serves several special purposes. Obviously it allows you to measure the supply current, but so would a supply-current meter or a current meter connected in the short-circuited output lead. The resistor is important simply to limit the instantaneous output current from the very large capacitors which are usually connected inside the supply, across its output terminals. Evidence of the existance of these capacitors can be found in the spark you will often get if you short the supply directly with a wire (be sure to set the current limit near the bottom of its range before attempting this).

Note that it is the existence of these capacitors, that motivates the use of a small series resistor in each supply line on your prototyping board, and additional bypass capacitors on the circuit side of it. Such resistors serve to limit the initial transient current peaks when the supplies are switched on, should a short exist on the board. If the short involves semiconductor junctions, they might survive with a series resistor to limit the current. They certainly may not otherwise!

Once the supplies are turned on and the circuit is connected, continue to involve the current-limit control in an active way:

- First, using your DVM connected to each voltage supplied to your board, in turn, lower the corresponding current limit gradually, noting the point at which the voltage begins to drop. Then raise the control a small fraction of its range above that. Repeat for all supplies. You are now ready to proceed with your **Experiment**. If your supply has a current meter, you can switch it to monitor current during subsequent operation, although with an adjustable current limit, the output voltage is often more informative.
- During operation, at those times when you suspect that supply current may have increased, repeat the supply-voltage measurement, and raise the supply-current limit enough to return to the normal voltage level as needed.

B2 OSCILLOSCOPE CALIBRATION

While there is a limited amount you can do about the integrity of your oscilloscope, except measure it, calibration is something you should check early and often, to the extent that is easily possible. The only tools assumed are the oscilloscope itself, a DVM, possibly a frequency counter and a power supply. As well, your oscilloscope often has a calibration signal available to assist.

Early on, it is a good idea to check the nature of the calibration signal, its waveform, amplitude and frequency, using the oscilloscope itself. You can check its frequency with a frequency counter if that is available. You can check its amplitude with a DVM, through a calculation (too detailed and context-dependent to detail here) which involves the process by which your DVM measures AC, and the detailed nature of the calibration waveform. Otherwise, you can use your oscilloscope (independent of its state of calibration) to estimate its properties (whose specifications are no doubt given in the oscilloscope's operating manual):

- Display the calibration signal, using your ×10 probe for connection, with the channel display set for direct coupling. Adjust the time-base triggering and vertical sensitivity for a pleasing display showing a small number of cycles (1 to 4) extending vertically over more than 50% of the screen.

- Now, remove the probe from the calibration signal, ground it (or leave it open if the result is the same), and adjust the channel vertical position so the trace is at some convenient reference line on the display graticule

- Now, reconnect your probe to the calibration signal, noting carefully its peak levels, its peak-to-peak value and the time of one period (or more, as convenient).

- Now, remove the probe and connect it across a supply whose output is being monitored by a DVM.

- Now, adjust the supply to display a voltage equal to some convenient attribute of one of the measures of the calibration signal you have taken, and read the DVM to find the actual value.

- Now, use your frequency counter, if available, to check the calibrator frequency, and thereby the time-scale calibration. Note that if you don't have a frequency meter, you can calibrate using the powerline frequency assuming it to be closely the nominal value (usually 60 Hz, but possibly 50 Hz). Be careful when accessing the power line. (Your probe end connected to a slightly-exposed blade of a slightly-removed power-line plug, is one way, *but* consult your Instructor.)

Note that if during the amplitude- and time-calibrating process, you find more than a few percent (perhaps 5) deviation from nominal, you should be concerned, and report that fact to your Instructor. On the short term, you may be able to compensate using the variable controls typically available on each of the vertical and horizontal amplifiers. Note that such "recalibration" is generally *not* a good idea unless you are desperate.

Having completed the basic calibration of one channel, switch, first, the time-axis, and then the voltage-axis switches to other ranges, verifying the fact that the results seem to be consistent, to the extent that you can evaluate small signals and many cycles. As well, switch probe ranges, if that facility is available, in order to check probe gain. Of course, repeat the process for the second channel.

There are often special applications in which the characteristics of the two channels must be closely matched, through the process here called "normalization", and described in Appendix B3. You could, of course, at this point, so compare and standardize your two channels using both probes connected to the calibration signal, if you wish. Normally, it is better however to operate both channels with the variable gain control in the calibrated position.

B3 OSCILLOSCOPE MEASUREMENT

Here, we will consider a few basic aspects of the process of making reliable oscilloscope measurements. They will be presented as separated labelled segments, the last being on **Normalization**.

• Use of Probes

It is highly recommended that you *always* use a ×10 probe as a compromise between a set of conflicting requirements. As well, ×100 and ×1 probes exist, as do switchable combinations. The ×100 probe offers an obvious high-voltage capability of little interest here, as well as increased impedance, but typically not ten times that of a ×10 probe. While a ×1 probe offers the physical convenience of a probe, that is a handle and so on, it tends to load the circuit being tested in potentially confusing ways. Even with a good ×1 probe, one having a resistive inner conductor and shield combination which reduces ringing and high-frequency pickup, observed results are often not "real". However, it is unarguable, that relative to a ×10 probe, a ×1 probe provides 20dB additional gain, nominally "for free", and sometimes this can pay for a lot of other aggravation. If probe loading is a problem (certainly with a ×1 probe, but possible even with a ×10 probe), include a series resistor between the probe tip and the tested node. Use as large a value as you need to reduce node loading, while being aware of its effects on the frequency response and voltage calibration of the probe. The pole introduced by the resistor is easily estimated assuming a value for probe input capacitance. Typically, for a ×10 probe the input capacitance is around 3 to 8 pF and for a ×1 probe it is more than ten times larger, and at least 50 pF!

• Use of External Triggering

Always organize the triggering mode of your scope quite consciously; If at all possible, **use external triggering**. Increasingly, automatic features of a modern oscilloscope make the triggering process so easy that there is a tendency to ignore this outstanding feature, one which changed the way oscilloscopes were used, forever, when introduced commercially by the founder of Tektronix shortly after World War II.

For virtually any **Experiment**, one of the first things you should connect is the oscilloscopes external trigger input port to the fixed-amplitude trigger source that a modern signal generator normally provides. Immediately thereafter, with your probe connected to the generator's main output, observe the waveform while adjusting the oscilloscope's, trigger polarity, level and sensitivity knobs for a good display. Fondle these knobs a bit while watching the oscilloscope react to your caress! It can be quite stimulating!! So much so, that even if your circuit doesn't need an external generator, you might try it anyway! To do so with a triangle waveform is really great! More seriously, external triggering is so fundamental to good measurement practice, that even if your circuit generates its own signals, immediately identify a special node to which an external trigger lead can be attached. Often you can use a ×10 probe on the trigger input to reduce circuit loading. Some modern oscilloscopes honour this possibility by allowing you to display the trigger input as a third channel.

An alternative, but one which can often confuse later issues, is to use channel A to display a reference signal from which you always trigger *internally*. (This idea makes me feel bad as I write it, and even worse to re-read it, but it is an otherwise popular approach.)

• Channel Normalization

While it is recommended, in general, that oscilloscope channels be operated at fixed gain (with the variable gain control in its reference position), and also with trace displays separated on the screen, there is an important exception: This is the case in which the application requires convenient direct comparison of two signals which differ in only some small way. Toward this end, the process here called **"normalization"** is often very appropriate. To **"normalize"**:

- First place the variable gain controls in the calibrated position;

Appendix B–5

- Second, chose the channel coupling appropriate to the final application;
- Third, chose a gain-switch position, the same for both channels, which is compatible with the smallest signal in the application, to the source of which signal two probes can be connected;
- Fourth, with the input probes both open, or (better) joined to ground, adjust channel position controls until the display traces coincide at the desired screen position (vertically);
- Fifth, with both probes connected to the same source, and two signals displayed, adjust the variable-gain control of the *higher-gain channel* until both traces precisely overlap. *At this point*, the channels can be said to be "normalized". Note that the process produces no guarantee of absolute accuracy, but only of comparability on only one range with only one coupling combination.

A few additional comments are useful: First, note that if the variable control must be changed a great deal from the calibrated position, it is a sure sign that your oscilloscope needs repair! Report this, to your Instructor, and try to obtain another one. Now, if you are satisfied with that issue (or sufficiently desperate), with the probes still connected, change the coupling to direct. Obviously the two displays should still overlap, but you possibly may require a change in *both* gain switches to allow the signal (with its dc component) to be visible on the screen. In any event, vary both gain-switch settings to verify that both behave in the same way. *At this point*, you have done all that is easily possible to make subsequent comparative measurements relatively painless. (See also, B4 (3) below.)

B4 DIFFERENTIAL MEASUREMENT

For the present purposes, differential measurement refers to the process of measurement of signals between two points of a circuit, neither of which is grounded, and both of which may include other signal components than the one of interest, at least dc, and often high frequencies. Moreover the impedance level of each node, separately, may be very high. A classic situation where differential measurement is needed, is the measurement of the signal across an ungrounded series circuit component (perhaps to evaluate the current flow through it). Various approaches to such measurements and their relative merits follow:

(1) The approach recommended generally in this **Manual**, where the situation normally allows it conveniently, is to make separate ground-based measurements and subtract the results, mentally or otherwise. Certainly this approach is acceptable if the difference is large, and its properties obvious, and/or if the individual measurements are very precise. Such is easily the case with DVM measurements for example, but also, there, the small size and possible battery operation of a DVM allows it to be used to measure such differences directly. For more about direct measurement using floating instruments in general, see (4) below. Direct measurement of differences is also quite straightforward with a modern digital-memory oscilloscope which has the feature of storing and subtracting waveforms acquired at different times. It is particularly convenient for the evaluation of the consequences of a series of changes of a circuit parameter on some subtle performance measure over a very long period, of hours or even days. The existence of this facility can be viewed as evidence of the legitimization and normalcy of the process of taking a sequence of single-probe measurements whose results are combined to affect a differential measurement.

(2) Some oscilloscopes, labelled as differntial input, have a special "front end" to which two probes are connected, which allows balanced differential measurements. But note that for various reasons discussed in *Section* 7.8 of the **Text**, the common-mode-rejection bandwidth of such a differential amplifier is much lower than that for the normal display mode.

(3) Many dual-trace or multiple-trace oscilloscopes provide a poor-man's differential capability by allowing a channel signal to be inverted, and the outputs of two channels to be added for display. Though this is a useful feature, bear in mind that the common-mode rejection that results is very limited as a result of channel gain and frequency response differences, and the complexity of the process. The situation can be improved if the two channels are "normalized" beforehand. Alternatively, both here and for "normalizations" in general, an alternative exists: It is to invert one channel with gain switches, probes and so on being the same, and while displaying a channel sum, connect both probes to one of the nodes to be measured, or otherwise to a very large signal of the offending kind. Now adjust the channel variable gain controls one at a time to identify the combination which makes the displayed signal smallest, and the two channels most equal.

(4) As noted above in the context of reference to a DVM for direct differential measurement, it is occasionally possible to place the instrument (electrically) between the two nodes, or across the series component. As noted, this is straightforward with a battery-operated instrument, since only the instrument body capacitance (and lead capacitances) provide a common-mode load on the circuit, and the (small) effect on each node is nearly the same. However, for line-operated instruments, such is *not* usually the case, since the power supply of such instruments (usually, but not always) provides a large capacitance to earth, or power-supply ground, of the internally common part of the instrument circuit. Even though this input connection is nominally isolated, it has a large capacitance through the power-supply capacitance, while the other (an amplifier input) has almost none. Worse still, the nature of line-operated instruments is such that its measuring terminals are typically not symmetric, one often being connected to the instrument case and thereby to power ground. Such is the case (!) with an oscilloscope for example. In such cases (!!) it is possible however to "float" the instrument for measuring very low frequency phenomena. But one *must be very* **careful**, for the process

Appendix B–7

involves isolating the instrument ground from that of the other instruments using a "grey" ground-isolation adapter in the connection between the instrument line cord and the electrical power outlet. If you contemplate such a measurement, *first*, do some preliminary checks with your oscilloscope physically isolated on the work bench. Proceed by connecting a DVM on a 200 V ac range between the case of the oscilloscope and that of another (ground-connected) instrument. Now remove the oscilloscope line-cord plug; insert the adapter; and read the voltage. *Be Careful!* The voltage may be large, typically half the line voltage, but sometimes larger. Now, carefully, evaluate its source impedance (and available current) by shunting the DVM by a resistor which reduces the DVM reading to one half its former value. At this point you know the power-frequency current which can flow in your circuit, *if your circuit has grounded parts*. If this level of current can be tolerated, you can proceed, *but carefully*. If your circuit has no grounded parts, as is the case if no external generators are used, and your regulated power supply, if used, has an isolated output, expect no problems. In the case of an isolated supply the only current which flows is through the capacitance (ideally small) of the isolating power transformer.

B5 AC MEASUREMENT

• Waveform Sensitivity

This sub-section will cover two issues. The first concerns the nature of DVM ac measurement: the second concerns ac oscilloscope measurement. Though your DVM is inevitably equipped with an ac range whose readings are nominally rms values, what the readings actually mean, in any but the simplest case of single-sine-wave measurement, can be an open question. Fortunately, it is relatively easy to discover what is actually going on! First, recall that the rms idea is related to the measurement of power. Specifically, the rms value of a waveform is its heating value. Instruments which measure rms values directly are traditionally relatively expensive, though modern electronics is beginning to provide alternatives involving, for example, computation. Thus the usual modest-cost DVM uses some other approach. All such approaches rely on the fact the most usual use of an ac meter is to measure the value (nominally rms) of a single sinusoid, at a single frequency, normally, that of the power line. The first approach of note is to measure the average value of the rectified input. Half-wave rectification is usual. The second approach is to measure the average-to-peak value of the input. Both approaches are increasingly easy with electronic circuits, though the former is more insensitive to impulse noise. In both cases, under the assumption that the input is a sine wave, the reading is calibrated to a corresponding rms value. In practice, using your waveform generator with sine-, square- and triangle-wave measurements, it is relatively easy to discover what your DVM actually measures. Independent of all this, your DVM is still useful for comparative amplitude measurements of a fixed wave shape, without concern for the technique used.

With respect to ac measurement using your oscilloscope, what can be said is relatively obvious. First, peak-to-peak measurement is typically easiest, particularly if the display vertical position is adjusted to make it so. Second, peak above (or below) average is easy to measure with the channel ac-coupled. For some waveforms, the rms value can be easily calculated.

• Coupling

A second attribute of DVM ac ranges concerns their behaviour at very-low frequencies. Just as with an oscilloscope for which that option normally is available by switching, the idea of ac measurement implies ac-coupling usually understood to be embodied in an (input) coupling capacitor which blocks dc. Thus, it is normal, but not necessary, to have such a capacitor in a DVM for ac measurement. When present, its size is chosen to guarantee reasonably accurate measurement at 60 Hz, with 3dB points of a typical DVM possibly at 20 Hz or less, and also at 20 kHz or more (for other reasons, of course).

B6 TIMING MEASUREMENT

The need for timing measurement is very basic in electronics. Although you may associate the idea and terminology to be exclusively digital, as is emphasized in *Section* 13.1 of the **Text**, it is much more general. What is often forgotten is that for linear circuits the related idea is called phase, simply because frequency measurements, and the implied ubiquitous sine wave, led electrical engineering in that direction in its early history. *Now*, in spite of the apparent divergence of these two traditions, they are the same really, or at least can be viewed as such! In each, one is interested in the time interval between successive occasions at which a signal passes a certain *reference value*. In linear circuits, normally seen as a sea of sine waves, the critical reference level is zero, the average value of an isolated sine wave, and the time measured is between zero crossings. Between successive zero crossings, the time is called the half period, and between every second one, it is called the period. And the period is very important in linear-circuit analysis, though its reciprocal often clouds the scene. Likewise, between two different sine-wave appearances, say on 2 nodes, it is the time between zero crossings in the same direction, that is important, but is typically expressed with respect to a standardized measure of the full cycle, stated as an angle, but, still, really a time in disguise!

For digital signals, there are various conventional reference levels. For measurement of the elapsed time between two different signals one often uses the gate threshold as reference, or approximates it by the half-way point in the signal swing, or alternatively and more straightforwardly, halfway between the supplies. Note well however, that as observed earlier, there must be a reference level used to define a time of occurrence at the heart of such a measurement.

For a single digital signal, as conceived in a nominal sense, there is the transition time, or, more confusingly, the rise time or fall time. The confusion is a semantic one, relating to different interpretations of the word "rise" (and, correspondingly, of the word "fall"). In general, "rise" implies "up", but also has a connotation of "start", through "arise".

Following the "up" idea, a signal going from low to high usually is said to rise. This is particularly appropriate using the *positive logic* notation, where going up implies a transition from logic "0" to logic "1". Unfortunately negative logic exists as well, to confuse the mind. Moreover the "start" aspect of the word leads to the idea that rise time is what happens first, no matter in what direction. Here, in this **Manual**, as in the **Text**, we generally use positive logic in which "rise" makes most sense, and, for safety, prefer the idea of using the word "transition" with a notation for which one, whether low-to-high or high-to-low!

Now the practical problem of transition-time measurement is where to begin and where to end. Conventionally, the 10% and 90% levels of the observed change are used as reference levels to define times. Occasionally, other definitions involve 10% of a supply voltage or of a nominal signal swing, etc. The reaons for the 10% choice is that it is often difficult in practice to define where the beginning and end of a signal transition actually are. (That is in a sense why propagation delays are more important, and more often used.) In practice though, one often sees an abrupt change at the beginning, in which case fastidious use of the 10% point is often foregone.

When actually measuring times, whether transition or propagation, on an oscilloscope screen, it is reasonably effective to try to align the signal appropriately with a convenient screen reference level. Use of the display vertical-position and variable-gain knobs is recommended. In very modern oscilloscopes with increasingly improved human-factors design, controllable cursors can be used to define reference levels from which times are often automatically measured.

In practice, the overall time of operation of a system, called the group delay in a linear (communications) world, or phase delay, or propagation delay, is probably best done by actual system test where, as Heisenberg suggests, invasive disturbing measurement is minimized. In this **Manual**, we see a dynamic (!) example of this idea in the (repeated) use of ring oscillators in **Experiments** #13 and #14, where, in an ironic reversion to the origins of Electrical Engineering, frequency is measured, and time is derived!

B7 LINEARITY AND DISTORTION

What follows will be a relatively brief comment on the relatively large topic of linearity and distortion, with the emphasis here on measurement:

To review, non-linear behaviour originates in the fact that the basic active devices of electronics are inherently nonlinear — junction devices being exponential (or logarithmic) and field-effect devices being square-law. For each, there is a small-enough signal such that operation can be perceived to be linear to any desired degree. For junction devices, junction voltage signals of $10n$ mV (for n the junction slope factor), and corresponding current changes of less than $50/n\%$, are seen to be acceptable, providing a total distortion of $9/n^2\%$ or so (see the **Text**, page 256). For field-effect devices biassed at $v_{GS} = 2V_t$, signals of amplitude $V_t/10$, providing about 20% current change and 5% distortion, are seen to be barely acceptable. (See the **Text** page 390.) As well, for very large signals, devices change regions of operation, going from one where the previous linearization tends to apply, to where it does not, to cutoff for example for either type of device, or to saturation for BJTs, or to triode-mode operation for MOS.

The result, variably called limiting, clipping, saturating or cutoff, typically occurs first at one signal peak, and then at the other. Now, the point of interest here in asymmetrical clipping is that while the original waveform had a zero average, the resulting one does not, implying a change of bias (that is of dc level) in associated circuits. It is important to note this because it leads to behaviour which often seems strange at first sight, particularly in capacitor-coupled circuits. Related phenomena occur, especially in BJT circuits, when coupling capacitors are used for signals large enough to cause rectification, and so-called dc restoration (See page 195 of the **Text**). However, these situations are too context-laden to pursue here at this time.

Another good approach in visualizing non-linear behaviour, particularly at low distortion levels, is by direct comparison of input and output waveforms on the oscilloscope screen. The technique is independent of waveform, with sine and triangle waves each offering their own advantages. The two waveforms can be overlapped vertically, to the extent distortion allows, by juggling the position and fixed and variable gain controls on the two channels. Exact overlap is possible if a "Delaying Sweep" facility is available on the oscilloscope for compensation of the relative phase shifts of the two signals.

Returning to emphasize minor non-linearities, with signal compression considered as its extreme form, but also minor clipping included to a degree, the best test waveform to use is a triangle wave, rather than a sine wave. The reason is that the rising (and falling) edges of the triangle, in bending, present "non-linearity" quite graphically. The only problem a triangle introduces is a need for sufficient bandwidth in the signal region of interest to sustain harmonic amplitudes and phases well enough to maintain the classic triangular shape. Certainly this is not problem for direct-coupled broadband amplifiers, but can be for more band-limited applications. When testing with a triangle wave, a very easy comparative measure available is the reduction of one peak relative to the expected triangle, constructed (mentally) by extending the slopes at the waveforms middle (ac "zero crossings"). Thus it is easy to see, (for example), that one signal has dropped 10%, while another has dropped only 5%, and reasonably possible as well to detect even a 1% or so drop as an indication of the onset of distortion.

B8 TEMPERATURE TESTING

The issue underlying the words "temperature testing" here is two-fold: Our goal is typically both to assess the temperature of a component, and to change it. We will say little about assessing it, beyond a comment on the use of a finger for such purposes! There are both good and bad techniques for "fingering" a component. In general, *be very careful*; learn to tentatively *caress* the component attempting to sense radiated heat before contact is made; only grasp a component likely to be hot, *after* you have done this preliminary test: Your thumb and forefinger are too precious to burn! As well, of course, there are other more sophisticated methodologies beyond our need here, including temperature-calibrated wax crayons, and of course thermistor probes.

On the topic of changing the temperature of a component, there is much more to be said: The potential application in mind is generally the assessment of the sensitivity of a particular circuit design to temperature differentials. A related but different issue is the sensitivity of a whole circuit to the ambient temperature to which the whole circuit is exposed uniformly. We will say little about the latter, except to point out the relevance of a hair dryer equipped with a diffuser made of non-flammable material (say a bag or wad of cotton cloth, but *be careful* of fire); use a low heat setting. Concerning the heating of individual components in relative isolation, a drinking straw through which one blows is a very handy device. For cooling, a drop of alcohol or ether on the component before blowing is quite useful, though recommended only with care (and permission of your Instructor). The only problem, but not a serious one usually, is the obvious change in humidity this blowing produces around the circuit. Be careful to avoid droplets of moisture. On an even more simple basis, even a finger on a component will often suffice for minor temperature changes, although the capacitance of the finger can introduce other confusing effects. The tip of a soldering iron held near the component, or in contact with it if you are not concerned much with its integrity, is a way to produce very large (possibly destructive) temperature changes. To avoid some of these problems, a surrogate heater in the form of piece of metal on an insulating handle (for example, the tip of a small screwdriver) can be used to transfer heat from a source such as a soldering iron, or even a flame (which is definitely *not* recommended *here*). *Be very careful* with the soldering iron, both for your own safety and for that of the prototyping-board plastic material.

Appendix B-12

B9   ROLES FOR POTENTIOMETERS

• What is a Potentiometer?

The term potentiometer, or *pot* for short, as used in this **Manual** and commonly (but not exclusively) elsewhere, refers to a three-terminal resistor, one having an adjustable connection (usually called a "*tap*") which can move to, or near to, the resistor ends. The term potentiometer originated in relation to the use of the third (tap) terminal to provide a potential intermediate between the *potential* of the end terminals. The use of the term **pot**, with reference to a tapped resistor, much predates the other rather-more-common street usage.

While a potentiometer is inherently a three-terminal device, two terminals only (the tap and one end) are often used to create a variable resistor. Good practice is generally to connect the other end to the tap, such that for a failure of the sliding mechanical parts (in which the internal slider no longer contacts the resistive element), the resulting resistance is limited to that of the end-to-end value of the resistive element.

• Physical Construction

Potentiometers are of many kinds, depending on material and mechanism. Common materials generally include wire (wound in a squared helix form), carbon film or a solid carbon composite, or metal film. Wire typically has the greatest power rating, but the available adjustment is quantized as the slider goes from wire to wire; Carbon film is continuous, but has a relatively-low power-handling ability and has relatively low stability; Carbon composition has a higher power rating and greater stability typically; Metal film is relatively high power, has continuous adjustment, is quite stable, and exhibits well-controlled end-connection behaviour. The latter comment is with respect to the way the ends of the resistor are connected, and whether, and by what means, the slider has access to the very end. Often the mechanical arrangement is such that the minimum resistance from tap to pot end undergoes a sudden jump in value as the slider just touches the end-connection.

While there are many different adjustment mechanisms available, a major attribute of them is whether they involve fraction-turn or multiple-turn adjustment. A further distinction, related to precision and repeatability, is whether the multiple-turn aspect is a property of the resistor itself or of only the slider-control mechanism. For example, elegant 10-turn "precision" wire-wound pots actually use a multiturn spiral helix of resistive wire!

In this **Manual** the preferred pot is a 10kΩ multiturn (22 or 25 turn) device, with a metallized (film) ceramic element. However, while this offers very high-resolution and setting stability, it is often tedious to adjust. If that is the case, a single-turn metal-film unit may be preferred. Very low-cost fraction-turn carbon film units are often adequate, and commonly used, but not very professional, nor well suited to a Laboratory setting.

• Areas and Techniques of Application

Generally-speaking, potentiometers are viewed by some with a degree of controversy: There is one school of thought which tends to scorn them as evidence of amateurism in design. Its devotees try to avoid pots totally, through cunning use of various kinds of feedback. Historically, however, they often have had to stoop to potentiometer controls for the *human interface*, such as is embodied in the controls on the front of your power supply. Increasingly, though in relatively higher-cost applications, computer-based-digital or touch-sensitive-continous controls are replacing mechanical potentiometers.

At the other end of the spectrum there are actually amateurs who tend to make every resistive element a variable one, as a (very poor) substitute for (even rudimentary) analysis. Clearly these folks are not engineers nor engineers-in-training, and should not concern us here. A middle view is that *one* potentiometer (or perhaps one per function) can be used to make up for all of the uncertainties in the circuit, not in the design (for that is assumed to be well-done), but in the components whose tolerances are large because their cost is low. Finally there is the view that potentiometers are quite legitmate as tools through which the sensitivity of a circuit to change of a particular parameter can be evaluated, and compensated. The two application directions, whether as tool or tolerance absorber, tend to lead to two different styles of use: As tool, one can often be more casual, since if one approach doesn't work well, another is often available. But, in a formal design, such luxuries rarely exist, and the designer has to do more initial work. Broadly speaking, the following generalizations often apply:

1) Use potentiometers in low-frequency parts of the circuit for measurement or compensation.

2) For high-frequency applications, you *may* be able to use a very *small* potentiometer, but stray capacitive effects, particularly during adjustment, are very troublesome, requiring at least insulated tools and a lot of care. If this is the case, use an active component (FET or diode) through which the high-frequency part of the circuit is under control of a pot operating at dc.

3) For a potentiometer connected as a variable resistor, usually include a fixed resistor (a *padding resistor* or *pad*) in series to control the minimum value of the combination. This technique is recommended to cope with several problems: the minimum resistance of a pot is often not well known (but it may be zero); there is often a discontinuity in resistance near the end of the adjustment range, and zero ohms is not normally a very useful element in circuit design.

4) For a pot connected as a variable resistor, where the overall resistance of the available pot is unnecessarily large, shunt resistors can be used between the terminals, to limit the maximum resistance which can appear there in normal operation, or if the slider opens.

5) For pots used as 3-terminal elements it is often good practice to use a resistor in series with one of the connections, usually at one end, but often at the tap. Series resistors are often appropriate where an adjustment is made in a high-frequency circuit, in which case the fixed reistor is connected at the "most active" end of the pot, with one end of the pot itself at ac ground if possible, and as large connecting lead as necessary (but as short as possible) in between.

6) When using a potentiometer in a measurement application, keep in mind that the pot setting is often easily found (immediately) by voltage measurements at its terminals, perhaps combined with other data. It is in this regard that a fixed resistor in series with a variable one is very handy (see Appendix B13), for example. Of course, with the power turned off and the pot removed, one or two ohmmeter readings can be very informative.

B10   ROLES FOR RESISTORS

Beyond the obvious need for resistors in virtually every[1] electronic circuit imaginable, there are special roles which resistors can play in an experimental environment, whether in an academic laboratory or in one whose goal is electronic-product development: *Series resistors* of relatively small value incorporated in your breadboard are actually very useful components:

(a) They can serve as markers of connections, links, etc, particularly if relatively rare and garish colour combinations are used, especially values in the 5% scale involving red and orange, such as 130, 160, 240, 360, and 360 ohms. In this respect, 22 and 33Ω are particularly nice, though somewhat more likely to appear in a more usual circuit role.

(b) In the limit, one can connect only one end of such a resistor as a node marker, or self-labelling test point.

(c) Series resistors are generally very useful in evaluating current flow, both statically and dynamically. Certainly a DVM reading across a series resistor, particularly if unit values (100Ω, 1K and so on) are used, gives a very quick idea of a branch current. However, at high frequencies such resistors can prove a bit troublesome, but typically can still be selected in value to be useful. Unfortunately, in general, if large enough to make measurement easy, they can affect high-frequency performance.

(d) Series resistors can often serve to limit peak currents under unusual conditions. Accordingly, such resistors have been used in this **Manual** to couple external generators to sensitive circuits such as CMOS gates. Such resistors also can be used to measure input current (as in (c)). In general, when connecting different parts or modules of a composite circuit, such resistors can help, particularly, if one part is a high-voltage-high-power one, and the other relatively weak and susceptible to large currents.

(e) While the negative effect of series resistors at high frequencies has been mentioned in (c), the same effect can be used to damp oscillations in circuits, particularly where particular parts can have a negative-resistance component. Thus 100Ω series resistors can be used in the MOS gate connections in power amplifiers.   A related mechanism is in effect when a small-valued resistor (100Ω to 100kΩ) is inserted in series with the tip of an oscilloscope probe, isolating the probe capacitance from the node, and filtering some very-high frequency components. The same idea occurs in an extreme sense with large resistors in series with DVM leads, for dc measurements taken in active circuits. Here the resistor decouples the circuit both from the meter-lead capacitance, but *also* the circuit from the meter as a source of spurious signals, in which situation, it acts as an antenna in the complex electromagnetic sea that surrounds us.

(f) A special case of some of the previous ideas is the use of a small resistor (and/or mini-choke) in series with the *ungrounded* power supply leads for particular sections of your circuit, as assembled on your prototyping board (PB), or, correspondingly, at the power input of each PB of a multiple-PB system. High-frequency capacitors to ground on both sides of the series component, with a tantalum on the load side as well, are important. The resistor value should be chosen as large as can be tolerated but typically such as to cause a voltage drop of only a few percent in the supplied voltage for normal currents. For example a 10Ω resistor in series with a 10V supply providing 10mA, entails only a 1% loss in supply voltage.

---

[1] But not all. Certainly in modern IC design the attempt is to reduce the number of explicit resistors and the total resistance.

B11 CONTROL OF PARASITIC OSCILLATION

All circuits when actually assembled incorporate a variety of elements which are not represented explicitly on the circuit diagram. Such elements, referred to collectively as *"parasitic elements"* or simply *"parasitics"* consist in general of resistances in connections, leakage resistances, inductances in leads, and capacitances both between components and from components and connections to ground. For our puposes, it is the capacitances which are of most concern. Generally speaking, here in this **Appendix**, here in this **Manual**, and commonly in the real world of electronics, the word *parasitic* refers to a stray capacitance.

Realize that the internal capacitances of an active device, say an FET, are parasitic elements as we use the term here. They just happen to be such important parts of an important active device having only three-terminals and a relatively fixed structure, that we can identify them and proceed to understand them *once* for use *each time* we use a FET in a circuit. Thus the normal use of the word *parasitic* comes to mean what is left, what we can't organize, and even what we can't even identify explicitly. Such for example is the case with the array components used in the **Experiments** in this **Manual**. If the capacitances of a 3-terminal device occupy a large part of our concern for frequency response in the **Text**, imagine all of the possible internal capacitance couplings within such an array. The mind boggles!! They certainly are *not specified* on a data sheet. The combinatorial problem they present is quite overwhelming, and it is simply not solved. Moreover, the relative positioning of the components within such an array is not generally known. However, we do know the pin connections and pin-to-in capacitances are likely to be very important. Actually, leads aside, the largest capacitances are probably from each device to the substrate. That is one of the reasons one grounds the substrate to eliminate a major parasitic coupling path.

Thus when one assembles a circuit in general, and particularly when using an array, a vast network of parasitic capacitive couplings exists. Traditionally, one tries to limit these by the use of intervening grounds which terminate electric fields which would otherwise extend from one component to another. This idea, referred to as "shielding", is simply a larger-scale generalized version of the effect of the substrate we see in an IC. On the scale of a printed-circuit board,, such a substrate is called a ground plane, the closer to which a component is, the *less* it couples to adjacent components, but the *more* it couples to ground. Thus the evil of component-to-component capacitive coupling is traded for the lesser evil of component-to-ground capacitance.

This is one of the reasons that redundant ground busing and gridding is important on your prototyping board (PB), as it forms an approximation to a "ground plane". Correspondingly, to mount your PB on a metal plate connected to ground, is a good idea. In an environment, deficient in ground, such as your prototyping board, it is possible to add extra ground in the form of "shields", grounded bits of metal, intervening between components which must be close, yet between which capacitive coupling is troublesome. Pieces of plastic-covered metal foil with a connecting wire are often useful in this regard. Even one or more grounded insulated wires quite close to each of the potentially-coupled components can help in the PB environment.

Why all the concern about parasitics? It is simply that they often cause trouble, the most pernicious form of which is **oscillation**. Very often amplifiers, particularly high-gain ones, oscillate when assembled. Complex multistage amplifiers are more likely to do this than simpler ones, but even a single transistor is not immune. Feedback amplifiers are particularly susceptible, even though carefully designed. The oscillations are typically at a relatively high frequency where the impedance of a small capacitance can be of the same order as that of the circuit's resistive elements. In that sense, lowering average resistance level in a circuit can help, but compensating troubles often arise. It is often true that the frequency of oscillation exceeds that of the available equipment, and is not seen explicitly but, rather, is just sensed: While the amplifier seems to work, its output levels, perhaps as measured by a DVM, seem to drift around, be noisy and so on. A relatively sure test is to place your hand near the circuit, and/or even to touch nodes selectively with a finger. Resulting change is a relatively sure sign of oscillation.

But what does one do about it? Here are some directions in which to proceed:

Appendix B–16

- **Wiring Style**

    - Use as much ground as possible on your PB. Certainly a ground bus on each side is important. (As well, ground the metal case of your PB if that is available.)

    - Bypass the supply buses well to ground with short-lead high-frequency capacitors, both to provide more ac ground, and to prevent coupling paths through the supply.

    - Pay attention to layout, organizing the amplifier input and output to be as far apart as possible. In general modularize, for example separating input and output stages.

    - Separate supply connections to modules with small series resistors, perhaps providing only 0.1V drop at the dc bias level, using high-frequency capacitors to ground on both ends.

    - Use an appropriate kind of PB wiring style (see Appendix A3).

    - In general the short cut-lead approach is best, but be careful, in this somewhat formal mode, of the proximity of components between which one or more active elements intervene. When using wire extenders on components in the cut-lead style (or otherwise) put the wire extension on the grounded or low impedance end of the component, using a very short connection at the active (radiating or receiving) end.

    - For the flying-lead style, try not to cross components from different parts of the circuit. Occasionally cut the leads of components whose antenna effect (for either radiating or receiving) may cause trouble.

- **Detecting the Cause**

    - Bear in mind that for parasitc oscillation, as for any problem in the real world, there is usually *no single cause*, only an unfortunate interaction of effects no one of which alone is bad enough to cause the trouble (or its correction to cure it). This is probably the major source of the sense of helplessness that such real problems (both here and in the broader world) create in the minds of anyone who seeks a simple single solution. You must identify a **set of** *potential causes* and reduce the effect of **each** of them.

    - With the probe of your scope connected to a relatively low-impedance circuit node, relatively near the output of the oscillating section, run your finger around near the circuit to detect the part which seems most sensitive.

    - Now with a short wire (say, a resistor lead) in your hand, touch *all* circuit nodes in the sensitive part of the circuit, noting the effect, if any. Nodes for which no effect is noticed are not in the oscillating loop or have too low an impedance to matter (and thus not likely to be receiving nodes in the parasitic-coupling process). Survey all nodes quickly before proceeding. Notate your circuit diagram with ↑, –, ↓ to indicate the effect.

    - Nodes for which the output increases or changes waveform dramatically, may be either a radiator, or a receiver, or simply a neutral participant. A small capacitor (one to hundreds of pF depending on circuit impedance levels) placed on the node to ground will help you to decide: Ideally the length of the lead connected to the node should be very short. For the frequencies we will encounter here, the other end can be of usual length or even extended (by a few centimeters, only) by an extra wire. If there is no change, the node is likely to be a relatively low-impedance radiator. If the output drops, the node may be a receiver or a high impedance radiator. Knowing the frequency and the capacitance, you can get an idea of impedance level to compare with what simple circuit analysis seems to indicate.

    - About now, you have data on all relevant nodes and rough ideas of the degrees of their sensitivity. Now, the challenge is to hypothesize causes, and then to test them. One general approach in testing them is to try to make the conjectured couplings *controllably worse*, as a means to verify your hypothesis. Typically, you join small capacitors (say 10pF) between nodes identified in your

Appendix B–17

scanning process which are either physically close (as connected on the board) or electrically close (as seen on the schematic).

- A second general approach in testing them is to overwhelm the effect of all parasitics using a shunt capacitor, initially very large. The idea is to introduce a dominant stabilizing pole in the parasitic loop. (One difficulty, incidentally, is that there are often several parasitic loops simultaneously active, so that two or more dominating capacitors may be necessary.) This is obviously often the case for amplifiers with explicit (external) feedbacks, in which the process can be viewed as a search for *the* dominant pole (parasitic or otherwise), often confused (as usual) by the fact that there may be more than one. Sometimes this process produces an amplifier which does not oscillate, but does not do the intended job either! However the result can be quite useful, since by lowering the stabilizing capacitor the circuit can be brought just beyond the edge of oscillation, where node probing can be done again, but with much greater sensitivity, in search of the "other" causes.

- Less of a diagnostic or testing tool, although of some use in both these roles, is a related idea, namely the use of compensating feedbacks introduced to cancel part of the effects of a parasitic coupling. The idea has an ancient tradition in early radio-frequency electronics, employing vacuum triodes, particularly those using tuned circuits, where it was called *neutralizing*. A related idea exists in modern circuits intended to generate negative resistances using the Miller effect with positive-gain amplifiers. In all of these cases, an explicit local positive-feedback connection to a node directly driven by one source (the undesirable one) (possibly through a parasitic) provides a compensating current which in the limit exactly cancels the undesirable original input, as though it was not there. In practice, exact cancellation is difficult and inevitable over-compensation can lead to negative impedances. Usually one is satisfied with cancelling a part (say 90%, in which case the effect to reduce the original parasitic coupling to 10% of its original value). Once you have identified a sensitive node to which an undesired coupling exists, the challenge is to find a source of the appropriate opposite polarity. The origin of the idea with tuned circuits was largely in the recognition that a tapped inductor, with tap grounded, provides the required antiphase voltages at its two ends. Duplicating the size of small parasitic coupling, is typically no problem, since capacitive voltage dividers are quite easy to create.

B12 STANDARD COMPONENT VALUES

- **Standard 5% Values with Marked $\overline{10\%}$ and $\underline{20\%}$ Values**

    $\underline{\overline{1.0}}$, 1.1, $\overline{1.2}$, 1.3, $\underline{\overline{1.5}}$, 1.6, $\overline{1.8}$, 2.0, $\underline{\overline{2.2}}$, 2.4, $\overline{2.7}$, 3.0, $\underline{\overline{3.3}}$,

    3.6, $\overline{3.9}$, 4.3, $\underline{\overline{4.7}}$, 5.1, $\overline{5.6}$, 6.2, $\underline{\overline{6.8}}$, 7.5, $\overline{8.2}$, 9.1, $\underline{\overline{(10)}}$.

- **Standard 1% Values with Marked "*Unit*" Values**

    $\underline{1.00}$, 1.02, 1.05, 1.07, 1.10, 1.13, 1.15, 1.18, 1.21, 1.24, 1.27, 1.30, 1.33, 1.37, 1.40, 1.43, 1.47, 1.50, 1.54, 1.58, 1.62, 1.65, 1.69, 1.74, 1.78, 1.82, 1.87, 1.91, 1.96, $\underline{2.00}$, 2.05, 2.10, 2.15, 2.21, 2.26, 2.32, 2.37, 2.43, 2.49, 2.55, 2.61, 2.67, 2.74, 2.80, 2.87, 2.94, $\underline{3.01}$, 3.09, 3.16, 3.24, 3.32, 3.40, 3.48, 3.57, 3.65, 3.74, 3.83, 3.92, $\underline{4.02}$, 4.12, 4.22, 4.32, 4.42, 4.53, 4.64, 4.75, 4.87, $\underline{4.99}$, 5.11, 5.23, 5.36, 5.49, 5.62, 5.76, 5.90, $\underline{6.04}$, 6.19, 6.34, 6.49, 6.65, 6.81, $\underline{6.98}$, 7.15, 7.32, 7.50, 7.68, 7.87, $\underline{8.06}$, 8.25, 8.45, 8.66, 8.87, $\underline{9.09}$, 9.31, 9.53, 9.76, $\underline{(10.0)}$.

- **"Unit" Values:**

    1, 2, 3, 4, 5, 6, 7, 8, 9, (10).

- **"Tens" Values (with {Geometric} and [Arithmetic] "Means"):**

    1, {3.16}, [4.99], (10).

- **Colour Coding:**

Colour	First/Second/Third Digit	Multiplier	Tolerance
Black	0	1	
Brown	1	10	1%
Red	2	10E2	2%
Orange	3	10E3	3%
Yellow	4	10E4	4%
Green	5	10E5	
Blue	6	10E6	
Violet	7	10E7	
Gray	8	10E8	
White	9	10E9	
Gold		0.1	5%
Silver		0.01	10%
No Colour			20%

Examples:
Resistor with four bands: brown, black, red and silver.

    Resistance = $10 \times 10E2 \, \Omega = 1000 \, \Omega$; Tolerance = 10%.

Resistor with five bands: brown, green, black, red, (space), brown.

    Resistance = $150 \times 10E2 \, \Omega = 15 \, k\Omega$; Tolerance = 1%.

APPENDIX C
Reporting

1	The Role of Engineering Reports	**181**	*4*	Report Formats	**185**
2	Report Design as Preparation	**183**	*5*	Standard Forms/Graphs	**189**
3	Engineering Record Keeping	**184**			

C1   THE ROLE OF ENGINEERING REPORTS

Those who attempt to capture and communicate the essence of Enginering as a discipline, have said that "Engineering is all about problem solving". And it is accepted that there is a certain generalized truth in that statement. But what are these problems? What is their nature, and that of the implied solutions? In response to these questions, it is quite natural for one to begin to think in physical terms: say in terms of popular problems of public policy today, about global warming, overpopulation, decaying roads, a circuit that doesn't work, and so on. And yet to do so is to miss a very important point. It is that these words do not truly capture the problems to be solved, but merely portray their most obvious symptoms. For problems take form and become real only in the minds of individuals and of groups of individuals in a society. Thus, it is that the physical origins of important ecological problems today lie in physical acts which once seemed, in earlier minds, quite ordinary and benign. It is in this sense that problems are simply perceptions. It is this reality which lies at the root of the apparent difficulty in solving the problems of society, today. For problems are perceived differently by different groups and individuals. It is in that sense that the problem itself is not real, but only its perception. The same goes as well for its solution. Correspondingly, the usual stumbling block in the path of any solution lies in its perception as relevant, worthwhile, and possible!

But what on earth does this have to do with Electronics? with Laboratories? with Reports? The common factor, I submit, is **communication**. It is apparent that a problem does not become universally real unless it is *communicated.* Likewise, a solution does not become acceptably real unless it is *communicated.* Yet, typically, they, problems and solutions, are each products of separate minds! Accordingly, the challenge we face, both in the Laboratory and at large, is to integrate them, and **communication** is the means.

Thus, this **Manual** has attempted to communicate a view of a relevant problem as I, and perhaps your Instructor, perceive it, *our* problem being to understand *your* perception of the real world of Electronics. If the **Manual** has begun to do its job, you too may be coming more aware of the need for the skills you are practicing here. But how do we know what you *now* know? How do we know what you perceive as the solution to our conjoined problem – the enhancements of *your* perceptions of Electronics. How does anyone know what an engineer perceives to be of relevance to their perceived problem. Well, in case you had not already guessed, a possible way is through the mechanism of an engineering report.

Engineering Reports can take many and diverse forms: They can, for example, be presented orally, perhaps informally, as in response to your Instructor's request for the status of your work, or formally, as a presentation to your class on *your* perceptions of the issues and outcomes of relevance. But of more direct concern to all of us here is the written kind of Engineering Report, of which, again, there are many varieties. But, as noted elsewhere, and often, they all, oral or written, share a set of common properties, of which the most important is clarity through structure, organization and relevance.

Concerning relevance, it is very important to know and understand the context of the report requested: What is really asked? What is really wanted? Thus when your Instructor asks "How is it going?", know that the request is not as casual as it sounds. Though she may graciously accept a casual "OK" in response, a more precise answer would always communicate more: "We have finished part x with the graph you wanted. Here it is. Would you like to see it? We are now setting up part y." The same idea surfaces when dealing with a more formal engineering report, usually in written form. What is its intent? Is it to present a broad overview, a so-called Executive Summary with general observations and conclusions, or is it to be a detail-loaded step-by-step description of the entire process, or should it be both, or neither? Thus often, as a young working engineer you will be asked simply to present relatively raw results, in as structured and tabulated form as you can conceive, graphed if possible.

Inevitably, they are needed quickly; inevitably they must be perceived as complete. If you are not sure what is *really* needed after a short period of intense reflection, and *some work*, then ask for clarification. Do so *early*, but *not often*. When in doubt, *do more than you are asked*, but remember the stated or implied time scale. In the real world, a perfect result too late can be worthless: Think of your boss. (S)he needs *your* work to answer he big boss' question at a formal meeting in two hours. Can he say his engineer isn't finished yet? Of course not! For to say so is to raise questions about his own management skills, and his willingness to tolerate incompetance around him. Your boss may ask for too much, but what he wants most is *something* (the best *you* can do in the given time), on time! It may not suit him, but at least he has something to work with.

Appendix C–3

C2 REPORT DESIGN AS PREPARATION

- **Table-Driven Experimentation**

In the performance of experiments in general, you must face the resolution of two conflicting goals: On the one hand, experimentation is about the dynamic process of acquisition of knowledge and experience. On the other hand it is also about a more prosaic process as well – the acquisition of data on which to base later reflection and anlaysis intended to lead to improved understanding. To satisfy the *first*, the process required is one of relatively rapid goal-directed measurement based on initial conjectures, and leading to more. To increase the rate of data gathering, measurements should be orderly, but rapid, with data held and processed dynamically, perhaps only in the mind. But, for the *second*, accurate record keeping of carefully-taken and thoroughly-checked measurements is essential.

In reality, in your career as engineer (or applied scientist), these two goals are hard to separate. Often, in hindsight, following a quick experiment to test your understanding of the situation, you will wish you had recorded enough data to allow you, with new insight, to test yet another conjecture, through analysis. Yet, you recorded very little of significance! You wish for better, but settle for less – and face the need to repeat the work! But there *was* an alternative, a way to observe quickly, but with some carryover.

Generally speaking, it is to prepare for the situation appropriately, *before doing any experiment*. In a formal sense, this process can be called *experimental design*, but it need not be *very* formal, to be *very* useful.

Broadly stated, the idea, is to run through the experiment in your mind before acting, reflecting on what is to be measured, what is to be expected. As you do this, in order to organize your thinking, and to help later, prepare a table in which to record expected results: Columns can be for a group of related measurements, while rows represent time sequence or parameter change. Label them both quite thoroughly. It is often appropriate to reserve blank rows and columns for inevitable hindsights, and intermediate calculations. Perhaps use every second row and column. A few extra columns at the right of your table are often handy. Sometimes making each table box large enough for 2 or 4 entries, allows for remeasurement, for the purposes of checking, increased precision, etc.

Now, finally, back to the original point, the conflict between a quick overview and a long-term record, we see that the table can help. As well as presenting the big picture, a projection of the totality of the experiment to be done, it represents a sample space in which to jot down, as quickly as will ever be possible, a snapshot of glimpsed results. Perhaps there is no time for more; but if time is found, the place already exists!

Appendix C–4

C3 ENGINEERING RECORD KEEPING

Each of you, as engineers-in-training, should acquire a (well-) bound notebook for dedicated use in your laboratory work, your *Lab Book*. This book is to be used as an on-going diary of your preparation, analysis and experimental work. It is equivalent to the workbook required of employee engineers by some corporations. Correspondingly, your Lab Book can be used by you in many situations, for example, in support of any oral examination you may face in the laboratory, during the preparation of formal reports as the need arises, and at term end in any laboratory-experiment-based laboratory test.

In your Lab Book, reserve the first few pages for a list of contents (to be filled in later as needed). Correspondingly, it is good form to number the pages. As well, each page and/or entry should be dated (and sometimes witnessed by your coworker or boss, in industry).

In general, the Lab Book should be diary-like, not pretty, not perfect, but **complete**, in time-order, and with data serial-accessible. If you accidentally, or otherwise, do work on the back of an envelope, or in some other disorganized fashion, then this should be taped in your book, or at least well-summarized. This, of course applies directly to access by one member of a laboratory party to notes taken by another in the heat of experimentation (and, naturally enough, recorded first in the originator's Lab Book)! Note, as well, to make most effective use of the tabular scheme recommended here, you must afix the tables to your Lab Book in some acceptable way. (Magic) transparent tape is normally a good approach, but, as usual, your Instructor will give you the final word on this topic.

Again, I repeat, a Lab Book is not a thing of beauty, but it should be *orderly*; it is not a substitute for understanding; but it is a means to understanding, and thus should be *complete*; if you must copy others' work into it, do so for a good reason – in particular, don't believe that a full book replaces, in any way, the work required for true comprehension!

It is important for you to know that the topic of Lab-Book style is the subject of some contention; not that the notebook itself is an issue, but rather its physical nature. Properly speaking, the reason that a working engineer or scientist uses a ruggedly-bound notebook is one of *integrity*, in all of the meanings of that word. It is simply that to document the ideas, the process of design, the process of test, and so on (should a question arise at a later time), is important to many employers of engineers. The underlying reason is a very good one: It is that such an engineer's work is typically important, and often essential to the success of his corporation. Thus it must be recorded, dated, witnessed, and retained, in a manner about which there can be no eventual question. But the issue of some contention is "How?". I personally prefer, and recommend, a spiral-bound book as an economic and effective kind of Lab Book. Obviously, as usual, your Instructor will give you the last word on this topic.

Appendix C–5

C4 REPORT FORMATS

• Overview

As noted often, earlier, reports can take on many different forms. Obviously, in the present context, your Instructor will tell you exactly what is needed.

Nevertheless, a few general things can be said:

First, it is very important to record the context in which the work was done. When, where, by whom, on what topic, on what circuit, with what equipment, with what objective? Obviously there are a great many stylistic options available in which to provide even this limited data; they range from entries in a preformed table, through cryptic allusions to memos and other instructions, to an essay style using well-formed sentences, paragraphs and so on. None of these is more-or-less correct, more-or-less important, for those value judgements depend on context, whether this is made explicit or not. You should be capable of any or all styles. What is most essential in any context of report preparation, is the possibility of complete information recovery, as needed.

Second, it is very important to relate the results as carefully as you can to specific aspects of the request for them. Thus independent of the detailed style, cross reference to the questions being asked, context described, purpose intended, should be included. Thus, specifically here, tables and graphs should be labelled, both with well-formed, informative phrases that convey context, and symbolic and mnemonic reference to sections, parts and task components within an **Experiment**. (See the comments on referencing to follow.)

Third, it is very important to relate conclusions carefully to the results that motivate them. Parenthetic remarks such as "as demonstrated by line x in table y", or "see table y, line x", or simply "$y: x$" are quite useful. But often, what has to be said parenthetically is far too big for the space available: To include it is to disrupt the flow; to omit it is to lose the context (if it is needed later). This dilemma is resolved easily by footnotes, or reference notes of some other kind, simply a letter or number which causes the reader to go to some standard place to learn more (if needed). Such is the role of **Note** and **Identification**(ID) entries in tables which you will shortly see.

Fourth, it is important to try to state your conclusions, both ongoing and final, as specifically as you can, for the very act of putting them into words can crystallize your thinking. Thus, a statement to the effect that having done the **Experiment**, you feel good and have a warm glow, is not really what is needed. Rather something more concrete is better. This may simply be a restatement of the validity of a conjecture; but don't just parrot the words in a formulistic sense. A statement about some aspect being verified, but some other part not verified, is more convincing. Correspondingly, a statement about the percentage deviation of some result from the expected one is more compelling than either "it works", or "it doesn't work", where both statements are laden with some unspecified assumption of tolerance or toleration.

• A Tabular Mechanism

As noted earlier in various contexts, it is important to structure, unify and interrelate the data from your experimental work, since such processing is more likely to reveal the patterns which underpin true understanding and expertise in any subject. It is this search for pattern and unifying principle, in recognition of their role in complexity resolution, which distinguishes engineering from other technical and professional areas where case study, precedent, and rote-learning tend to dominate historically, and slow the present pace. It is this which makes engineers most likely to cope in a rapidly changing technical world, and is, one might argue, at the heart of the very pace of technological change itself!

The **Table** formats suggested minimally in the many **tabulation** items in each Exploration are a concrete attempt to proceed in the overall direction of table unification. In each **tabulation**, you will note that the somewhat standardized (but cryptic) pattern includes the item "a,b,c, etc" followed by a string of "wheres", "fors", and so on. While the detail of the presentation may leave something to be desired, the idea is quite important: Generally speaking, and with exceptions, both intended and in error, the list a,b,c and so on, is ordered with the items listed first tending to represent one of two things: either conditions related to a circuit topology, (such as a resistor-value or connection), or a running variable, such as the frequency at which measurements are to be made, again a sort of condition on the circuit, but externally imposed.

Generally speaking, each table sketched in **tabulation** can be viewed in isolation, as a small structure for local use, or, more grandly, as part of the big picture presented by all or part of the entire **Exploration**, of which it is a part. Note that the general problem, which the large size of many **Tables** attempts to address, is the multidimensionality of the small but real world of experimentation: There are many parameters that can be controlled, all of which can affect one or more of the other parameters to be measured. The difficulty is that the written page confines us to only two dimensions straightforwardly. The rest must be encoded some way. Thus the pattern which emerges is one in which the **Table** can be seen as a succession of smaller tables stacked one under the next (in the inverted gravitational field in which ice cubes find themselves in your drink), where each layer represents access to a new dimension of the bigger picture.

Now several attributes of the **Table** emerge immediately: *First*, it can be very wide, in fact too wide to fit across the page! Obviously such wide tables can be turned physically on the page if you like. But there is a better way, which is to turn them conceptually, as we will shortly see. It is fortunate in this respect that tables are rarely both wide and deep or, rather, that they must be so! Prototype **Tables** in two possible formats are to be found at the very end of this **Manual** for your use as models or directly by photocopying.

Second, it is possibly apparent that an entire **Table**, though large and with very many boxes, need not be filled out completely. This follows for several reasons including the fact that large parts of it, particularly at the conceptual left, are structural, with a particular parameter changed only once at the beginning of a part of the **Table**. As well, there is the immediate fact that generally speaking, the **Explorations** do not ask directly for *all* measurements possible, but only a sample.[1] Furthermore there are cases in which some measurements for some conditions are too straightforward to follow up, being obvious in some way. The net result, in summary, is that the **Table** may be relatively empty. But its major virtue is that if you want to measure something, immediately or in hindsight, there is potentially somewhere to put it, rather than it being lost in a maze of rough notes, comments, etc.

Third is the fact that the size of the **Table**, its inherent complexity, and that it can include the results of several parts of one **Exploration**, can lead to confusion as to where the data is, or alternatively, where it is commented upon. To counter this tendency, you could obviously use a lot of little tables, placed in your notebook in the sequence in which you took the data. Such tables are quite straightforward responses to a **tabulation** item, and you are welcome to use them, of course only as your Instructor approves. But let me assure you that the larger **Tables** have advantages that should be sustained, and include another mechanism to solve this problem, as outlined now:

Fourth is the fact that the **Table** includes a variety of notational aids, notably the Note, Label, and Identification (ID) columns and rows, and the Notes Sections at the bottom or right side of the table body. (Look at the Table prototypes presented at the end of this Appendix, to better appreciate these comments.) The intention of these elements is multifold: In general, they provide a location for *pointers* as a way to connect the table entries to the context of both the **Explorations** instructions and to the textual parts of your report. As well, the space at the bottom of horizontally-oriented tables to which pointers can be directed, allows for immediate notes on context, origin of instructions, significance of results, location of further commentary, and so on. Incidentally, when referring to the **Measurement**, or **Analysis** parts of an **Explorations**, a convenient reference notation to use is a combination of the **Exploration** number with the item sequence number as counted within an **Exploration** division. Thus, for example, E2.3 Mb) represents step b) in the 3rd part of **Exploration** E2.0. Similarly, in responding experimentally to **Analysis** items, use, for example, E2.3 A2, to notate a relationship to the second **Analysis** item in E2.3.

Fifth is the fact that the **Tables** include *Part* and *Table* boxes as well as a *Title* box whose role it is to extend the size of the table as needed, to make the horizontal one wider and the vertical one deeper.

---

[1] You of course are welcome to do more if time or inclination suits, with the **Table** format being a way to extend some of the **Experiments**, somewhat combinatorially.

Sixth is the fact that through the use of dotted lines, the **Tables** can serve two other purposes as well: For those who are comfortable with a larger writing style, a whole box (with solid borders) can be used for each entry. For those willing and able to write smaller, several options exist: Obviously, one can double the size of either **Table** vertically in a straightforward way. As well, it is obviously possible to use only the top of each solid box, reserving the lower half for other options, like remeasurement, correction etc. Far less obvious, and perhaps too complex, is that the width of both tables can be doubled, in effect, by using each solid box for two separate variable entries, one above the other, and labelled using the Symbol and Unit boxes creatively.

Seventh is the possibility of creating your own tables by photocopying parts of the prototypes provided, either directly or with photoreduction. Alternatively, creating a Post-it ® Pad, of small table forms, might be a useful idea.

• Presentation of Waveform-Timing Information

In the body of this **Manual** there are numerous occasions in which the presentation of waveform information is required, often motivated by the word *sketch*. For a single waveform, the word is relatively appropriate. However, for multiple related waveforms, the term diagram is more indicative of the need for the structure and order needed to communicate detailed timing (or phase) relationships. For complex situations it is often very worthwhile to do this on graph paper (for which a prototype is provided in Appendix C5), although in using graph paper, there is an implied tendency toward perfection which is typically not warranted. Of course, your Instructor will provide you with detailed advice in this regard.

Often better, in my view, than a perfectly-scaled aligned multiwaveform chart, is a scheme of notated (possibly)-separated waveform sketches. The sketches can be placed for example on a circuit diagram near the node to which they apply. Using this style, waveform sketches can often be created first somewhat in isolation from the general picture. Thus if one knows that there is a sine wave propagating through a circuit, one can *first* provide a sketch of a sine wave near a particular node, and then as *second* and *third* steps, label its voltage levels and timing, respectively. Though there is a tendancy to sketch only a single sine wave, there are many reasons why two cycles is better (for example, it provides more space for labelling). The only problem is that while drawing one sine wave cycle can be tricky, matching two is often quite a challenge (but a tracing template helps!). The idea of following an orderly process applies as well within each of the labelling tasks. For example an appropriate order established by someone biased toward ac amplifiers is based on the idea that peak or peak-to-peak values of the sine wave are most important, following which the average value, then absolute peak values can be considered.

Note that someone whose first concern is bias design would reverse the order of the first two. Likewise, with respect to timing, zero-crossing timing is typically (but not always) most important, with peak timing of secondary concern. Now what about relative timing? Once again, for relative timing between waveforms at relatively low frequencies, in midband so to speak, there is a priority to consider. In the limit, detailed phase doesn't matter. All that one is concerned with is relative polarity, that at a particular time while one waveform goes positive, another goes negative. From this viewpoint alone, one can now label relative time using some notation. An obvious one is to use t_0, t_1, t_2 etc, somewhat arbitrarily, or 0, $T/2$, T, $3T/2$, $2T$, etc, with direct reference to a clock period $T = 1/f$. For reasons which will become more evident when we discuss digital timing notations, in a moment, I personally prefer circled numbers 1, 2, 3 etc, which can be seen in use in a digital world for example in another ancillary to Sedra and Smith, "KC's Problems and Solutions".[2] I will describe what you see in more detail shortly. Now on all parts of such a composite diagram, all times marked 1 are the same, and 2 always, follows. You can view the numbers in many ways, as cycle (or half cycle) counts for example. If at some stage your concern extends to an interest in relative phase shift you can use a prime notation outside the circled number as a crude mnemonic for phase shift. Thus 3' can be seen to lag by a small amount, while 3''' suffers even more delay. On the other hand '3 leads 3, but lags '''3 in a corresponding way. Note again the emphasis on incrementality. You only do *what you need,* and typically only *when you need it!*

---

[2] KC's Problems and Solutions for Microelectronic Cicuits, 4/e by K.C. Smith, Oxford University Press, see page 366, for an example of the use of digital timing notation.

Contrast the overhead and work aspect of this approach with the setup necessary for a regular timing diagram!

The idea works very well in documenting digital circuits with a few added wrinkles. *First*, in logic, the prime notation has a more direct interpretation, as the delay through an average logic gate. Thus a signal entering a string of 2 inverters at time 4 emerges at 4″, two logic delays later. This idea is used to a considerable extend in "KC's Problems and Solutions". A second option, uses consecutive numbers to notate the output of a sequence of gates. In that scheme, a signal going into a gate at time 1 would emerge at time 2 to enter another from which it emerges at time 3, and so on. This scheme is handy for notating ring-oscillator waveforms where the numbers are cyclic modulo $2n$ where n is the number of gates in the ring. Another general approach is to label the input signal with interesting transitions at times 1, 11, 21,- - -, 101 and so on, in such a way as to allow intervening events to be labelled so as to indicate their relative time of occurrence. Thus 16 would appear more-or-less midway between times 11 and 21, while 10 appears shortly before 11. Of course, in each of these situations the prime notation can still be used to indicate average gate delays. It is possible also to assign a number of primes to the unit interval to reduce the need for long prime strings. Otherwise use a roman notation with primes indicated using the symbols i, ii, iii, iv, v, vi, etc.

In the event that cause and effect must be noted very distinctly, one can add the directed-arrow notation that is used to indicate timing sequence in some digital timing charts, using such an arrow on the target waveform with a circled number at its tail end to indicate the cause, both its relative time, and origin (as the waveform transition associated with the same circled number).

Appendix C–9

C5 STANDARD FORMS/GRAPHS

Finally, the four pages following contain forms intended for your convenience through photocopying. In order of appearance:

1. A 4-by-6 large-format linear graph.
2. A 7-decade log-linear graph.
3. A data table of the type referred to here as "*horizontal*", for 12 variable entries and 9 (or 18) sets of conditions.
4. A data table of the type referred to here as "*vertical*", for 17 (or 34) variable entries and 5 sets of conditions.

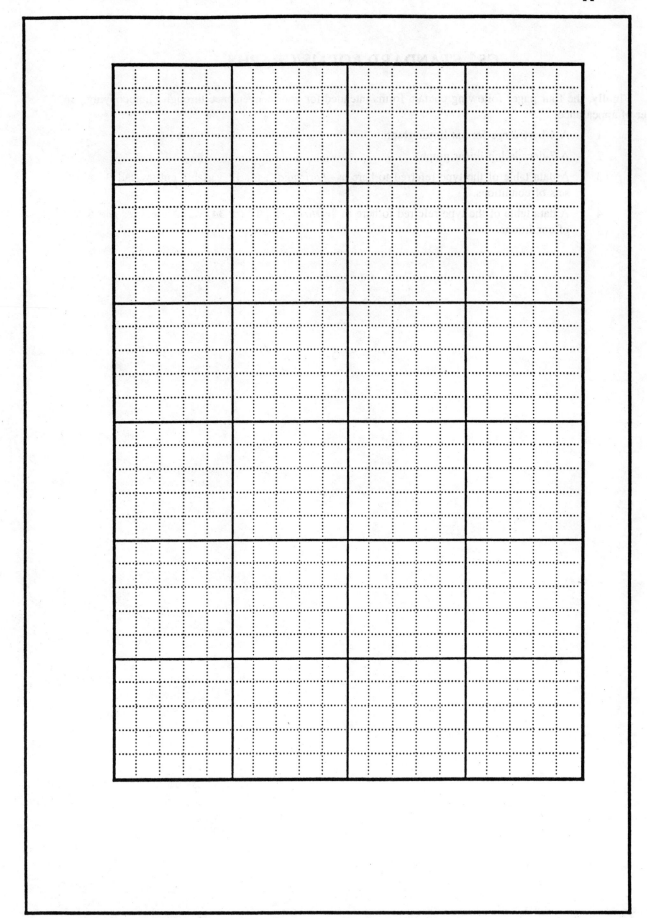

Appendix C-11

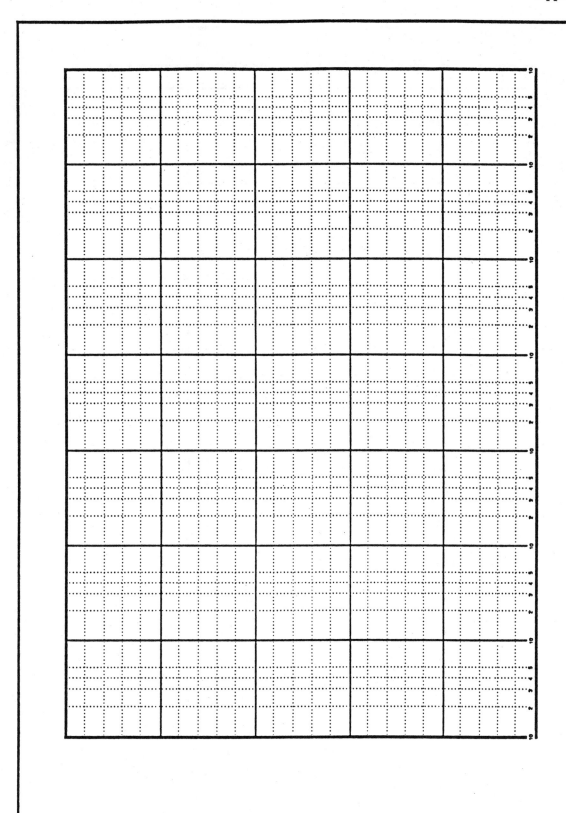

Appendix C–12

Exper	Part	Table	Title								Initial	Date
Note →												
Identification ↓ Label → Note ↓												
Symbol →												
Unit →												
a_1												
a_2												
b_1												
b_2												
c_1												
c_2												
d_1												
d_2												
e_1												
e_2												
f_1												
f_2												
g_1												
g_2												
h_1												
h_2												
i_1												
i_2												
Note#	Explanation, Extension, Interpretation											

Exper	Part	Table	Title:						Initial	Date

Note	Parameter Label	Symbol	Note+ID / Unit	a	b	c	d	e	Ref	Notes / Comment
	Parameter Label	Symbol	Unit						Ref	

Appendix C–13